An introduction to nuclear reactor theory

To Ruth

An introduction to nuclear reactor theory

C. E. Iliffe OBE, MA, CEng, FIMechE
Formerly Head of Central Technical Services
UK Atomic Energy Authority, Risley, England

Manchester University Press

Published by
Manchester University Press
Oxford Road, Manchester M13 9PL, UK
and
51 Washington Street, Dover
New Hampshire 03820, USA

British Library cataloguing in publication data
Iliffe, C. E.
An introduction to nuclear reactor theory.
1. Nuclear reactors
I. Title
621.48'3 TK 9202

ISBN 0–7190–0953–7

Library of Congress cataloging in publication data
Iliffe, C. E.
An introduction to nuclear reactor theory.
Includes index.
1. Nuclear reactors. I. Title.
TK9202.I38 1983 621.48'3 83–16227
ISBN 0–7190–0953–7

Typeset by Graphicraft Typesetters Limited
Printed in Great Britain by
Butler & Tanner Ltd, Frome and London

Contents

Preface

The following pages have been written in the hope that they will be of use to those who have come into the business of design and development of nuclear power stations from industries and disciplines outside this field. The aim has been to provide an introduction to the subject in the strictest sense of the word, that is, not to present a simplified version of the excellent text-books that already exist but to bring the reader up to a stage at which works of that kind may perhaps be more profitably studied. As far as possible an historical approach has been adopted so that the reader is confronted with the struggles of the early pioneers rather than being asked to jump to conclusions or to take things for granted. As befits a course of introduction, the reader is presumed to have no knowledge of nuclear power nor even of the fundamental physical principles on which it is based. Furthermore, the standard of mathematics required has been deliberately restricted to that typically attained by a graduate engineer.

The treatment starts with the earliest concepts of atomic structure and from these it attempts to trace an historical sequence of achievements, part experimental, part theoretical, leading to the realisation of nuclear power. The intention has been to limit the steps in this sequence to those thought to be essential for providing a continuous chain of reasoning from beginning to end. With so many parallel developments recorded at any one time, it is clear that an account of this kind cannot be unique. It is hoped, however, that the account that follows not only gives a sense of logical progress to the final goal but also includes the most notable contributions that were made on the way.

In the account given of these contributions, some liberties have been taken in the interest of abridgement and, in some cases, conversion to a more modern idiom, but the aim has been to combine historical accuracy with ease of understanding. At the same time, topics that are relevant to the present day have been treated in some depth in the hope of providing a firm basis for any subsequent study of more immediately applicable techniques.

It goes without saying that the information needed to reach these objectives already exists, but not between two covers. Excellent accounts of the early history of atomic physics are given in S. Glasstone's *Sourcebook on Atomic Energy*, in F. Soddy's *The Story of Atomic Energy* and in S. Tolansky's *Introduction to Atomic Physics*, and the first eight chapters of the present work are heavily indebted to them. Even in drawing on these, however, some parts of the material have been treated in greater detail whilst others have been only touched on or omitted altogether, depending on the extent to which they contribute to the purpose in hand. Inevitably, as the account comes nearer to the present day and goes into greater detail, less use is made of general works of reference and more of individual papers to learned societies.

Apart from some minor subsequent amendments, the present work was completed at the end of 1976 and copies of the typescript deposited in the libraries of the U.K. Atomic Energy Authority, and it is to this Authority that the author wishes to record his thanks for their permission and subsequent encouragement to undertake the work and to seek its publication. He also wishes to thank all those who have offered their valuable help and encouragement, either directly or through their colleagues, especially Mr. B. Cutts of U.K.A.E.A. Risley, Prof. W. B. Hall and Mr. M. J. Harris of Manchester University, Mr. J. L. Rowlands of U.K.A.E.A. Winfrith, Mr. S. E. Lewis of National Nuclear Corporation, Risley, and Mr. K. M. Leigh of U.K.A.E.A. Culcheth. The author also wishes to record his indebtedness to all his colleagues in the Central Technical Services organisation of the U.K.A.E.A. at Risley who have given him such a thorough education during his time with them, and in this regard he recalls with particular gratitude the patience and understanding of the late J. Erskine Mann. He is also grateful for the valuable assistance he received from Mr. Roland Smith and his staff in the U.K.A.E.A. Library at Risley, particularly in obtaining so many of the less accessible works of reference. Finally, he wishes to record his thanks to others of his colleagues at Risley, especially to Mrs. J. Jones for preparing the typescript of the first seventeen chapters and to Miss G. Bebbington for preparation of 66 of the 73 diagrams. He also wishes to thank Mrs. O. G. Lee and Miss D. Durward for preparing the typescript of the remaining chapters and Mr. Brian Hallett of U.K.A.E.A. Winfrith for preparation of the remaining diagrams.

C. E. Iliffe
Woodbury Salterton, Devon
29 January 1983

1

The atom in chemistry

1.1 The atomic theory of the nineteenth century

Although the more recent atomic discoveries have been made in the realm of physics, it was in that of chemistry that the first scientific evidence of the existence of atoms appeared. Chief among those to set chemistry on a rational basis was Robert Boyle, who, in 1661, clearly defined chemical elements as the undecomposable constituents of matter and allowed that many might remain to be discovered. He also clearly distinguished between chemical compounds and mixtures. Subsequent workers distinguished between different gases and indicated that gases have weight. These discoveries led A. L. Lavoisier (1743–94) to make accurate weighings of the reactants before and after the combustion process and thereby establish the Law of Indestructibility of Matter. By the year 1799 Proust had shown that the composition of a chemical compound was invariable.

Four years later, from the results so far obtained, John Dalton deduced his atomic theory, by which matter was said to be made up of small indivisible particles called 'atoms'. These atoms could be neither created nor destroyed, but only combined or separated. All atoms of the same element were attributed with the same weight, which was different from that of the atoms of any other element. From this theory, Dalton concluded that an element that combined with a fixed weight of another to form more than one compound, would do so in weights bearing a ratio to one another expressible in whole numbers. He assumed that these whole numbers were small and subsequently verified his view by means of experiment, thereby establishing the Law of Multiple Proportions.

The next step was that of assigning relative weights to the atoms of different elements. To determine the relative weights in which different elements combine was comparatively simple. The difficulty arose in deciding the numbers of atoms of each element in a molecule of the compound. Dalton assumed that these numbers would be the lowest possible, so that many of the relative atomic weights he deduced were wrong. To the oxides of carbon, for instance, he gave the correct formulae CO and CO_2, but to water he assigned the incorrect one HO.

These difficulties were partly resolved with the discovery by J. L. Gay-Lussac in 1808 of his Law of Combining Volumes. By this law, gases chemically combine in volumes that bear a ratio to one another, and to that of the compound if gaseous, expressible in small whole numbers. Three years later, Amadeo Avogadro interpreted this result by means of his hypothesis which attributes to equal volumes of all gases at the same temperature and pressure equal numbers of molecules. Thus the density of a gaseous compound at once gave its relative molecular weight, and this in turn, when multiplied by the fraction by weight of any constituent element, gave the product of the relative atomic weight of the element and the number of atoms of it in the molecule. By calculating this product for the same element in different gaseous compounds, and assuming that its lowest value corresponded to only one atom of the element in the molecule, the relative atomic weight of the element could be determined. In 1818, from considerations such as these, J. J. Berzelius was able to publish the correct relative atomic weights for elements forming volatile compounds.

A year later, P. L. Dulong and A. T. Petit had sufficient experimental evidence to conclude that the product of specific heat and relative atomic weight was the same for all metals. By this time, therefore, the relative atomic weights of most elements could be correctly determined, but it was not until 1858, when Stanislao Canizzaro brilliantly reviewed them, that their use, or even the atomic theory itself, became generally accepted. Until recently, atomic weights were quoted relative to that of naturally occurring oxygen as 16.0000. However, as will be seen in Section 3.4, the standard to which they are referred has since changed, but the consequent difference in numerical values is only some 4 parts in 100,000. Atomic weights on the current basis are given for the naturally occurring elements in the fourth column of the table in Appendix 2.

Further evidence for the molecular concept of matter was now found in the kinetic theory of gases, which was developed in the years 1850 to 1860, notably by R. Clausius and J. Clerk Maxwell. According to this theory, the molecules of a gas were assumed to be an assembly of free moving particles. On applying the laws of dynamics to this system, results were obtained that were consistent with Boyle's and Charles's gas laws and Avogadro's hypothesis.

1.2 The association of atoms and electricity

To the growing belief that matter consisted of atoms came now to be added the further idea that these atoms were somehow associated with definite quantities of electricity. This idea owed its origin to Michael Faraday, and was a direct consequence of the fundamental investigations he made into the nature of electrolysis in the years 1831 to 1834. Experimentally, he found that the amount of an element liberated at an electrode was proportional to the quantity of electricity passed through the electrolyte and that, for the same quantity of electricity, different elements were liberated in amounts proportional to the relative weights in which they chemically combine. The number of parts by weight of an element that combines with 8 parts by weight of naturally occurring oxygen is termed the 'chemical equivalent'. A mass of an element equal to its chemical equivalent was then found to be liberated by a quantity of electricity that was termed the 'Faraday constant' F and measured in units of coulombs per gram equivalent.

Faraday interpreted this result as being due to decomposition of the molecules of the electrolyte into positively and negatively charged submolecules, called 'ions', which were attracted to the negative and positive electrodes respectively, there being liberated and at the same time giving up their charge to effect the passage of the electric current. As Faraday clearly appreciated, the charge on an ion is associated with the atoms of which it is composed. For any element, the charge associated with one of its atoms in an ion must be proportional to the ratio of atomic weight to chemical equivalent. This ratio is termed the 'valency' and is twice the number of oxygen atoms with which one atom of the element chemically combines. The smallest value for the valency is 1 and elements having this value are said to be 'monovalent'. Each atom of a monovalent element liberated by electrolysis is thus associated with a passage of the smallest, or 'ionic' charge. For any other element the valency is expressible as a simple integral multiple of the smallest, so that the charge associated with an atom of it in an ion must likewise be a simple integral multiple of the ionic charge.

1.3 The Periodic Law

With closer approach to completeness and accuracy in the table of atomic weights, a striking feature of it began to emerge which not only provided an invaluable check on the table itself but was later to assist in revealing the very structure of the atom. The first to notice this feature was B. de Chancourtois who, in 1862, found that the elements could be marked off in order of increasing atomic weight along a spiral so that the start and finish of one complete turn about the axis corresponded to elements bearing some resemblance in their physical and chemical properties. Two years later, and quite independently, J. A. R. Newlands made similar observations which he published as a Law of Octaves. In a sequence of all elements in ascending order of atomic weight, the law stated that every element showed properties analogous to those of the eighth element beyond it. However, some of these analogies were difficult to accept, a drawback only partly due to errors in atomic weights and failure to allow for elements as yet unknown. At the same time Lothar Meyer expressed similar views and in 1869 he developed them further in conjunction with D. I. Mendeleyev.

The turning-point came in 1871 when Mendeleyev published the main features of what is now known as the Periodic Law. This law was the first of its kind to embrace all the elements. It was expressed in the form of a 'periodic table' in which the elements were arranged in successive rows, or 'periods', in order of increasing atomic weight and in columns, or 'groups', such that each contained elements of analogous properties which were systematically graded from one group to the next. The inconsistencies remaining in the table prevented its general acceptance, but they were dismissed by Mendeleyev either as gross errors in the prevailing atomic weights due to incorrect valencies or as failure to discover certain elements. The gaps in the table thus provided striking clues to the atomic weights and properties of elements that might still wait to be identified. Three of the clues led quickly to the discovery of gallium in 1875, scandium in 1879 and germanium in 1886. With this spectacular vindication the Periodic Law began to attract serious attention, not only as a pointer to undiscovered elements but as the final arbiter of atomic weight.

Although the principles of Mendeleyev's periodic table have remained unchanged, it has undergone some amendment since it first appeared. It is shown here (Table 1.1) in its present-day form. The elements are ordinally numbered from 1 to 98. The ten elements shown in parentheses appear to be too unstable to exist in nature to any appreciable extent. The elements of group 0 have properties that strictly place them between groups VIIB and IA, but as they were not discovered before 1895 they were grouped as shown to avoid a complete rearrangement of the table.

Table 1.1 Periodic table of the elements

Group / Subgroup / Period	I		II		III		IV		V		VI		VII		VIII	0
	A	B	A	B	A	B	A	B	A	B	A	B	A	B		
1	1 H															2 He
2	3 Li		4 Be			5 B		6 C		7 N		8 0		9 F		10 Ne
3	11 Na		12 Mg			13 Al		14 Si		15 P		16 S		17 Cl		18 A
4	19 K	29 Cu	20 Ca	30 Zn	21 Sc	31 Ga	22 Ti	32 Ge	23 V	33 As	24 Cr	34 Se	25 Mn	35 Br	26 27 28 Fe Co Ni	36 Kr
5	37 Rb	47 Ag	38 Sr	48 Cd	39 Y	49 In	40 Zr	50 Sn	41 Nb	51 Sb	42 Mo	52 Te	43 (Tc)	53 I	44 45 46 Ru Rh Pd	54 Xe

6	55 Cs 79 Au	56 Ba 80 Hg	57–71 Lanthanide series 81 Tl	72 Hf 82 Pb	73 Ta 83 Bi	74 W 84 Po	75 Re 85 (At)	76 77 78 Os Ir Pt	86 Rn
7	87 (Fr)	88 Ra	89– Actinide series						

Lanthanide series	57 La	58 Ce	59 Pr	60 Nd	61 (Pm)	62 Sm	63 Eu	64 Gd	65 Tb	66 Dy	67 Ho	68 Er	69 Tm	70 Yb	71 Lu
Actinide series	89 Ac	90 Th	91 Pa	92 U	93 (Np)	94 (Pu)	95 (Am)	96 (Cm)	97 (Bk)	98 (Cf)					

2

The electron

2.1 The electron as a cathode ray particle

In 1838, Faraday first turned his attention to the passage of electricity through low pressure gases, but it was not until 1854, when H. Geissler developed an improved vacuum pump and the technique of sealing metal wires into glass tubes, that really significant results could be achieved. Application of a potential difference between a pair of electrodes sealed into an evacuated glass tube was seen to produce a glow within the tube. Between the years 1879 and 1885, however, Sir William Crookes achieved the high degree of evacuation at which all glow within the tube disappears whilst a patch of greenish fluorescence appears on the glass wall opposite the cathode. By placing a metallic screen between this wall and the cathode, Crookes caused a shadow to be cast on the wall and so concluded that the fluorescence was produced by 'cathode' rays' travelling in straight lines from the cathode. Deflection of the rays by a magnet led him to the further conclusion that they consisted of a stream of negatively charged particles.

These conclusions received quantitative support from the classic experiments performed by Sir J. J. Thomson in 1897. Thomson's apparatus is illustrated diagramatically in Figure 2.1. The cathode rays emanated from the cathode C, passed through a hole in the anode A and were formed into a narrow beam by the slit S. The magnet M produced a magnetic field of flux density B at right angles to the beam so that the cathode ray particles, supposedly of mass m, carrying charge e and moving with velocity v, were deflected into a circular path of radius r, which could be determined from the position of the luminous spot on the scale D.

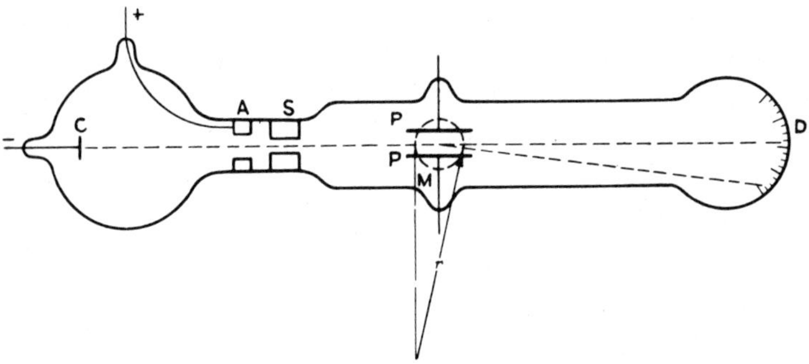

Figure 2.1 Thomson's apparatus for determining the charge-to-mass ratio of cathode rays

Equating the magnetic and centrifugal forces on a particle gave the relation

$$Bev = mv^2/r$$

whence

$$e/m = v/Br$$

The above expression for the force on a charged particle in a magnetic field, as well as that which follows for the force on it in an electric field, is the same whether the c.g.s. electromagnetic, c.g.s. electrostatic or M.K.S. system of units is used.

By applying a potential difference across the plates P, an electric field of strength X was then produced at right angles to the magnetic one and to the beam so that the particles were restored to their former straight line path. Equating the magnetic and electric forces on a particle gave the relation

$$Bev = Xe$$

whence

$$v = X/B$$

The ratio e/m could thus be determined from the known flux density B, field strength X and radius r. In this way Thomson found the velocity of the particle to be about one tenth of the velocity of light and obtained an estimate for the ratio e/m that was low by a factor between 2 and 3. The currently accepted value of e/m is

$$1.7588 \times 10^8 \text{ coulombs/g}$$

Thomson obtained the same results for different cathode materials and with different gases in the discharge tube, so that he was led to conclude that the cathode ray particles, or 'electrons' as they are now called, are essential constituents of all atoms.

2.2 Identification of photo-ions as electrons

The next step was clearly that of measuring the electronic charge e. Although Thomson was unable to make this measurement on the cathode ray particles themselves, other sources of negatively charged particles were already known as well as the technique for measuring the charge on them. The problem thus resolved itself into that of identifying, if at all possible, the particles from one of these other sources with electrons. One such source was provided by the photo-electric effect, which was first recognised in 1887 from the discovery by H. Hertz that a spark passed more easily across a spark-gap on which ultra-violet light was falling. A year later Hallwachs observed that irradiation with ultraviolet light caused a negatively charged metal to lose its charge but had no effect on a positively charged body. It thus became clear that negatively charged ions were emitted from certain metals on irradiation with ultra-violet light. These ions were known as 'photo-ions' and the process by which they were emitted as 'the photo-electric effect'.

In 1899, P. Lenard measured the ratio of charge e to mass m for photo-ions in an apparatus similar to that employed by Thomson for electrons, and illustrated diagrammatically in Figure 2.2. The cathode

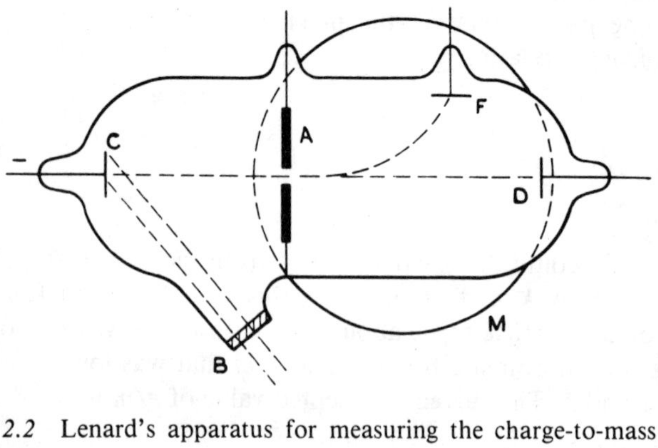

Figure 2.2 Lenard's apparatus for measuring the charge-to-mass ratio of photo-ions

C was an aluminium plate, struck obliquely by ultra-violet light entering the discharge tube through a quartz window B. The anode A was earthed and the hole through it made very small. A beam of photo-ions was thus caused to pass from the cathode, and through the anode, to fall upon an electrode D and there be detected with an electrometer. A magnet M could apply a magnetic field of flux density B at right angles to the beam, embracing the whole of its path from the anode to the electrode D. With this field applied, the ions traversed a circular path of radius r to fall upon a second detecting electrode F. Due to the potential difference P between cathode and anode, the photo-ions acquired a velocity v at the anode, given by the relationship

$$e/m = v^2/2P$$

As in the case of Thomson's experiment,

$$e/m = v/Br$$

so that the flux density B, potential difference P, and radius r being known, the velocity v and ratio e/m could be determined. Lenard found that the ratio e/m was the same as that for the electron, and he was thus able to identify the photo-ions with electrons.

2.3 Determination of the charge on a photo-ion by the cloud chamber

The determination of the electronic charge had become a matter of measuring the charge carried by a photo-ion. An apparatus for making this measurement was the cloud chamber, the principles of which had been discovered by C.T.R. Wilson in 1896. A diagrammatic representation of a Wilson cloud chamber is given in Figure 2.3. A cylinder A with a moving piston B is closed by a glass plate W. The top of the piston is covered with blotting paper moistened with indian ink, so that the air in the cylinder above the piston is saturated with water vapour. The inner surface of the cylinder is black. The piston is maintained in position by the pressure of the air below it. By means of a valve C the space below the piston may be connected to a vacuum chamber so that the piston suddenly falls, thereby producing an adiabatic expansion of the moist air above it. The air is thus cooled to become supersaturated, the excess vapour condensing into fine droplets in the form of a cloud or mist that is made visible by the light entering through the window L. Wilson found, however, that if the air

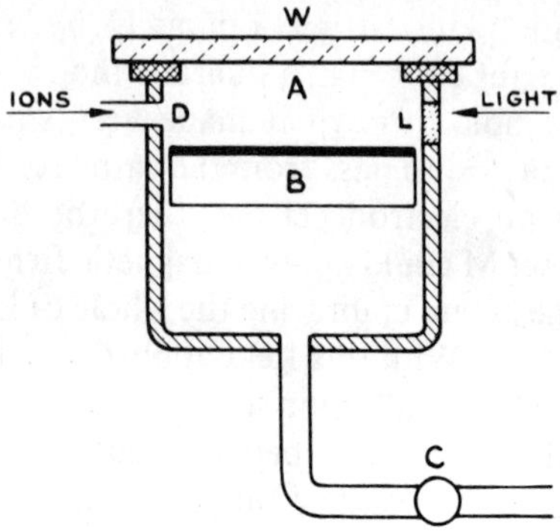

Figure 2.3 Wilson cloud chamber

in the cylinder was first filtered of all dust particles, no cloud appeared unless the expansion ratio was considerably increased. The dust particles thus appeared to facilitate cloud formation by acting as condensation nuclei. Wilson now showed that the expansion ratio required for cloud formation with dust-free air could be materially reduced on causing ions to enter the cylinder by the tube D. He therefore concluded that, in this case, the individual ions were acting as condensation nuclei.

A determination of the electric charge on a photo-ion by means of the Wilson cloud chamber was made by Thomson in 1899 using a method developed by J. S. Townsend two years previously. In this experiment a condensation cloud was precipitated upon photo-ions. The cloud was observed to fall slowly under gravity and the downward velocity of the top of it measured. By Stokes's formula, the terminal velocity of a drop of radius a and density ϱ falling in a viscous medium of coefficient of viscosity μ, is

$$v = 2g\varrho a^2/9\mu \qquad (1)$$

Identifying v in this formula with the downward velocity of the top of the cloud, and giving ϱ and μ values appropriate to water and air respectively, the radius a of the drops could be calculated. The expansion ratio r used to produce the cloud being known, application of the gas laws gave the absolute temperature at the end of the expansion as

$$T_2 = T_1/r^{\gamma-1}$$

where γ is the ratio of the specific heat at constant pressure to that at constant volume for air, and T_1 the absolute temperature at the start of the expansion. The mass of the drops in the cloud was thus

$$Q\tfrac{4}{3}\pi a^3\varrho = V_1\,(\varrho_{v_1} - r\varrho_{v_2})$$

where Q is the total number of drops, V_1 the known volume of air at the start of expansion and ϱ_{v_1}, ϱ_{v_2} the vapour densities corresponding to temperatures T_1, T_2 respectively. The quantities ϱ_{v_1}, ϱ_{v_2} being readily found from steam tables, this formula enabled the total number of drops within the cloud to be calculated. The total charge carried by the cloud was measured by allowing it to settle on to a surface connected to an electrometer, so that the charge on each drop, and therewith the charge on each photo-ion, was finally obtained.

2.4 Determination of the electronic charge by the oil-drop method

Although historically important, Thomson's determination of the electronic charge was subject to many inaccuracies. A very precise method of measuring the electronic charge was devised by R. A. Millikan in 1911, and his apparatus is illustrated diagrammatically in Figure 2.4. A pair of flat circular plates of metal P and Q were set accurately horizontal and parallel by means of a glass separator R. The plates were supported in a closed vessel V containing air at low pressure, and could be connected to the poles of a high voltage battery B. Small holes H were drilled in the upper plate P near its centre. These holes enabled an occasional drop of a non-volatile oil from a spray atomiser A to enter the space between the plates. Due to friction in the atomiser, this drop was electrically charged. With the battery

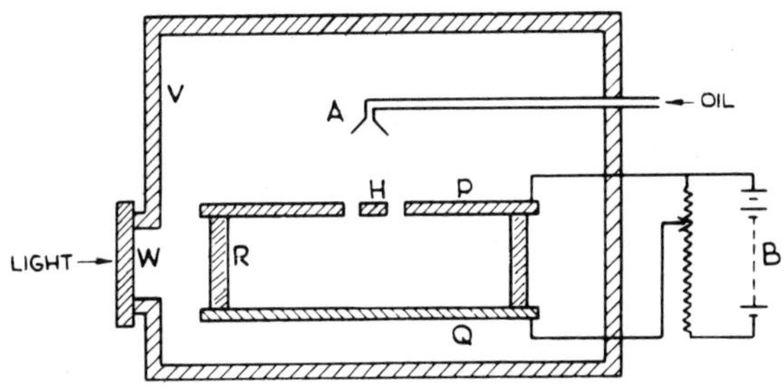

Figure 2.4 Millikan's apparatus for determining the electronic charge

disconnected, however, it would fall under the action of gravity with a terminal velocity v. Due to a powerful beam of light entering the window W, the drop appeared as a bright star against a dark background so that the velocity v could be measured by means of a microscope having a scale in the eyepiece. The radius a was thus obtained from equation (1), with ϱ and μ given values appropriate to the oil and air respectively, and the mass m of the drop found from the relation

$$m = (4/3)\pi a^3 \varrho$$

The battery was then switched on to produce a potential difference of about 1000 volts between plates P and Q. The resulting electric field of strength X caused the drop to move upwards with a velocity v' such that

$$v'/v = (Xe_n - mg)/mg$$

where e_n is the electric charge carried by the drop. Since v, m and X were known, measurement of v' enabled the charge e_n to be determined. From a large number of determinations, Millikan found the charge e_n to be always an integral multiple of a definite elementary charge, identifiable with the electronic charge e obtained by Thomson. Support was thereby given to a pronouncement by Thomson that negative electrification, though produced by very different means, is made up of units, each equal to the electronic charge. The accepted value of the electronic charge, as given by the oil-drop method, is

$$e = 1.6022 \times 10^{-19} \text{ coulombs.}$$

2.5 The Avogadro number

These investigations into the nature of the electron led to the inescapable conclusion that the electronic charge was identical in magnitude to the ionic charge associated with a monovalent atom liberated by electrolysis. This conclusion at once gave the number of ionised atoms in a mass of a monovalent element equal to its chemical equivalent as

$$F/e$$

Since the chemical equivalent of a monovalent element is equal to its atomic weight, the number of atoms in unit mass of such an element must be

$$F/e\mathscr{A}$$

when $\mathscr{A}$ denotes atomic weight. But the number of atoms in unit mass of any element can only vary with atomic weight and must do so in inverse proportion. It follows that this expression is not confined to monovalent elements but is quite general. The ratio F/e is written N_0, which is known as the 'Avogadro number', and is the number of atoms in a mass of an element equal to its atomic weight.

The units of F may equally well be written as coulombs per gram atom of a monovalent element so that the numerical value of F is seen to be affected by the recent change in standard to which atomic weights are referred, as noted in Section 1.1. On the present basis the current value of F is

96 487 coulombs/monovalent g atom

which, together with the value of e given above, leads to the result

$$N_0 = 6.0222 \times 10^{23} \text{ atoms/g atom}$$

The mass of the hydrogen atom is seen to be

$$1.0080/N_0$$

whilst that of the electron is

$$\frac{e}{e/m}$$

The ratio of the two masses is thus readily determined as 1837.

3

The mass of the atom

3.1 Positive rays

At first sight the mass of an atom would appear to be merely the reciprocal of the number of atoms per unit mass, which can readily be found for any element as described in Section 2.5. By measuring the masses of individual atoms, however, the result obtained in this way was found to be correct for only a few elements and to be in error for most of them. This discovery led to highly accurate methods of weighing an atom, which later enabled deductions to be made as to the energy that could be released from it.

Direct weighing of the atom arose from the natural desire to find a positively charged particle corresponding to the negatively charged electron. In 1886, E. Goldstein had noticed that if the cathode of a discharge tube were perforated, luminous rays emerged in straight lines from the holes on the side further from the anode. These rays were originally known as 'canal rays'. By introducing a hollow metallic vessel, or 'Faraday cylinder', into the discharge tube at the end remote from the anode, J. Perrin was able, in 1895, to collect these rays in it. An electrometer connected to the Faraday cylinder then showed the rays to be associated with a positive charge. This finding was confirmed in 1898 by W. Wien from observation of the deflection of the rays in electric and magnetic fields. Consequently, in 1907, Thomson gave these rays the more appropriate name 'positive rays', and it is by this name that they are now known.

3.2 Parabola method of positive ray analysis

The positive rays were further investigated in 1911 by Thomson, who devised an apparatus for measuring the ratio of the charge E to the mass $\mathscr{M}$ of the particles of which the rays were thought to be composed. This apparatus is shown diagrammatically in Figure 3.1. The positive rays are created in a discharge tube in the form of a flask F, fitted with an anode A and a perforated cathode C. Gas leaks slowly into the flask through the capillary tube B, the pressure in the flask being maintained fairly constant by adjusting the rate of pumping from the outlet G. The positive rays then emerge from the cathode into the observation chamber O. In order to avoid collisions between the positive ray particles and thus ensure they have a straight line path, the deflection of which can be observed, the observation chamber must be highly evacuated. The more highly evacuated is the flask F, however, the weaker is the beam of positive rays found to be. Thomson overcame this difficulty by maintaining such a pressure in the flask F that a sufficiently intense beam of positive rays passed through the cathode C, and by limiting the perforation of the cathode to a single fine bore hole so that a much higher degree of evacuation could be maintained in the observation chamber O by continuous pumping from the outlet E.

Having thus obtained a sufficiently intense beam of positive rays in a highly evacuated observation chamber, Thomson next set out to determine the charge-to-mass ratio by observing the effect on the beam of electric and magnetic fields. As his method for the corresponding determination in the case of cathode rays was only applicable to particles of equal velocity, it had to be altered somewhat for application to positive rays. As in Thomson's earlier experiments on cathode rays, an electric field could be applied by the plates P and a magnetic field by the magnet M, both fields being in directions at right angles to that of the beam, but the magnetic field, instead of being at right angles to the electric field, was now in line with it. The deflected rays were finally recorded on the photographic plate D.

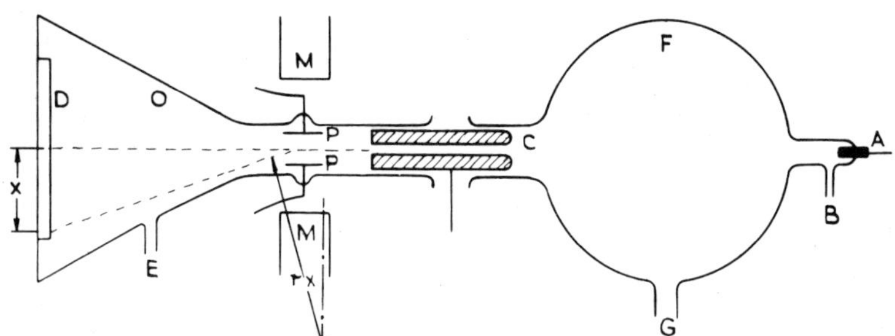

Figure 3.1 Thomson's apparatus for analysing positive rays by the parabola method

When neither the electric nor the magnetic field was applied, the positive rays formed a central undeflected spot on the photographic plate. On applying an electric field only, of strength X, positive ray particles of velocity v would be deflected into a circular path of radius r_x. Equating the electric and centrifugal forces on a particle gives the relation

$$XE = \mathcal{M} v^2/r_x$$

With the electric field removed and a magnetic field of flux density B applied, the same particles would be deflected into a circular path of radius r_y, the plane containing this path being at right angles to the direction of the magnetic field. On equating the magnetic and centrifugal forces on a particle, the relation

$$BEv = \mathcal{M} v^2/r_y$$

is obtained. Denoting by x the distance from the undeflected spot on the photographic plate to the spot formed by application of the electric field and by y the corresponding distance to the spot formed by application of the magnetic field, then since

$$x \propto 1/r_x$$
$$y \propto 1/r_y$$

it follows that

$$y^2/x \propto (B^2/X)(E/\mathcal{M})$$

the constant of proportionality depending on the geometrical arrangement of the apparatus. With given electric and magnetic fields applied together therefore, all particles having the same ratio of charge to mass (whatever their velocity) would form spots on the photographic plate tracing a unique parabolic curve, and from that the charge-to-mass ratio could be determined.

Thomson made his first experiments with hydrogen in the flask F and obtained a parabolic curve such as a–a in Figure 3.2. The axes of the parabola were found by first reversing the magnetic field to obtain the curve a′–a′ and then the electric field to obtain the curve a″–a″. The continuity of the curve showed that particles of all velocities over a certain range were present in the beam. In this way, the ratio $E/\mathcal{M}$ was found to be exactly the same as Faraday's ratio of a quantity of electricity to the mass of hydrogen liberated by it in electrolysis. The positive rays in hydrogen were thus recognised as a stream of particles, each of which was identical to the ionised hydrogen atom of electrolysis, having one electron less than the complete atom.

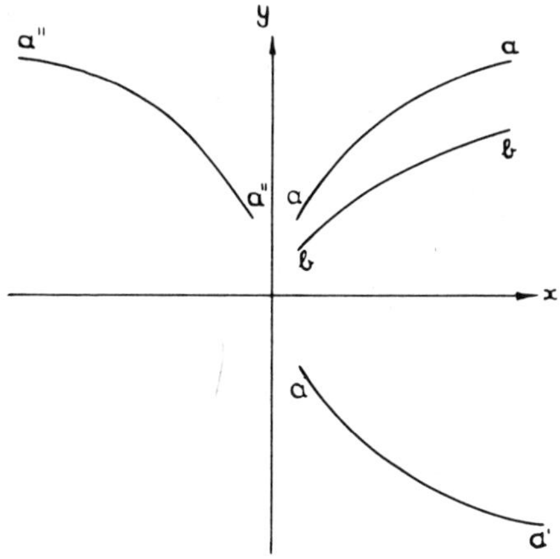

Figure 3.2 Form of record obtained by Thomson's parabola method

When the experiment was repeated with a gas other than hydrogen, the photographic plate revealed a curve, such as b–b in Figure 3.2, defined by a smaller value of y at the same value of x. For this gas, therefore, the ratio $E/\mathscr{M}$ was smaller than for hydrogen. The assumption that the charge E was the same for both gases then led to the relation

$$\mathscr{M}_b/\mathscr{M}_a = (y_a/y_b)^2$$

the two values of y being measured at the same value of x and the suffixes a and b denoting respectively hydrogen and the gas used in the repeated experiment. The ratio $\mathscr{M}_b/\mathscr{M}_a$ at once gave the atomic weight of the gas used in the repeated experiment and Thomson was able to establish that, with few exceptions, the atomic weights obtained in this way for various gases agreed with the accepted values.

Thomson came across one of these exceptions in 1912 in the form of neon gas, of atomic weight 20.2. He found that the photographs always revealed, in addition to the anticipated neon line, a fainter line corresponding to an atomic weight of about 22. No element of such atomic weight was known to exist. Since 22 is half the molecular weight of carbon dioxide, however, a possible explanation was that the fainter line corresponded to a carbon dioxide molecule containing two electrons less than the complete molecule. Some support for this explanation was provided by a faint line corresponding to an atomic weight of 44, but on purifying the neon gas, this line disappeared whilst that corresponding to the atomic weight 22 remained unaffected. Thomson next suggested that the gas used in his experiment might be a mixture of neon and a compound NeH_2, although no evidence of the existence of such a compound was hitherto known.

3.3 The mass spectrograph

To resolve Thomson's difficulty it became important to refine the parabola method in order to decide with certainty the precise atomic weights to be attributed to the two lines given by neon gas. The necessary refinement was provided by the 'mass spectrograph', invented by F. W. Aston in 1919. It was capable of measurement to an accuracy of one part in a thousand. Aston's mass spectrograph is shown diagrammatically in Figure 3.3. As in Thomson's parabola method, the discharge tube was in the form of a flask F. The positive rays passed through the bored cathode C to traverse the slit S. Positive ray particles of the appropriate velocity range were then deflected into a circular path of radius r_x by an electric field of strength X between the curved plates, P, and were thus enabled to pass through the pierced diaphragm H. The beam of positive rays emerging from this diaphragm was then deflected into a circular path of radius r_y by the magnetic field of flux density B due to the magnet M. This magnetic field was such as to deflect the rays in a direction opposite to that of their deflection by the electric field and thus cause them to fall on plate D. As in the parabola method, the high vacuum required in the observation chamber was maintained by continuous pumping from the outlet E.

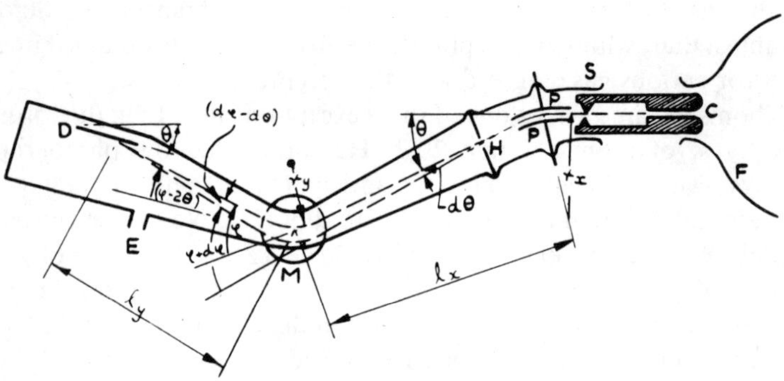

Figure 3.3 Aston's mass spectrograph

As in Section 3.2, equating the electric and centrifugal forces on a particle of charge E, mass $\mathscr{M}$ and velocity v gives the relation

$$XE = \mathscr{M}\, v^2/r_x$$

whilst equilibrium between the magnetic and centrifugal forces requires that

$$BEv = \mathscr{M}\, v^2/r_y$$

If θ, φ denote the angles through which the rays were deflected by the electric and magnetic fields respectively, then since

$$\theta \propto 1/r_x$$

and

$$\varphi \propto 1/r_y$$

it follows that

$$\varphi \propto B\sqrt{\theta/X}\,\sqrt{E/\mathscr{M}} \qquad (2)$$

Since θ was fixed, the value of φ determined the magnitude of the charge-to-mass ratio $E/\mathscr{M}$. The determination of the number of ionic charges on the positive ray particle was always inferential, but did not in fact lead to any error in the value of $\mathscr{M}$.

It would at first appear that the record obtained with the mass spectrograph, for particles of the same charge-to-mass ratio, would be only a small portion of the curve given by the parabola method. However, by an ingenious system of focusing, Aston was able to concentrate particles of slightly varying velocity on to the photographic plate and thus obtain a more well defined record. Due to variations in the particle velocity v, the beam of positive rays emerging from the plates P would be slightly divergent, making extreme angles with the cathode axis of θ and $\theta + \mathrm{d}\theta$.

At the magnet M, distant l_x from the plates P, the particles travelling in direction θ would be deflected through the angle φ, and those in direction $\theta + \mathrm{d}\theta$ through the angle $\varphi + \mathrm{d}\varphi$. At a distance l_y beyond the magnet M, the width of the beam would then be

$$b = l_x\,\mathrm{d}\theta + l_y\,(\mathrm{d}\theta - \mathrm{d}\varphi)$$

From equation (2), however,

$$\varphi^2 \propto \theta$$

so that

$$\mathrm{d}\varphi/\mathrm{d}\theta = \varphi/2\theta$$

and

$$b = \mathrm{d}\theta \left[l_x - l_y \left(\frac{\varphi}{2\theta} - 1 \right) \right]$$

The beam would be brought to a focus when b became zero, that is, when

$$2\theta l_x = (\varphi - 2\theta)\, l_y$$

Whatever the value of φ and therewith $E/\mathscr{M}$, it follows that the beam would be brought to a focus at the constant height $2\theta l_x$ above a plane through the axis of the magnet M having an inclination to the cathode axis equal and opposite to that of the beam leaving the plates P. The photographic plate D was therefore placed at this height, as shown in Figure 3.3.

3.4 **Isotopes**

With the mass spectrograph Aston was able to confirm the existence of two chemically indistinguishable forms of neon, one having an atomic weight of 22 and the other a corresponding weight of exactly 20, and not 20.2 as Thomson had supposed. It now became clear that the atomic weight of neon of 20.2 was the result of a mixture of some nine atoms of weight 20 to one of weight 22. This feature of neon was one which Frederick Soddy had, in 1913, suggested might be found in many other elements, and this suggestion was soon shown to be true, though not, at first, of oxygen.

In fact, by the end of 1920, Aston had examined nineteen elements in his mass spectrograph and found that nine of them were each a mixture of atoms of two or more different atomic weights very close to whole numbers. Furthermore, each element examined, for which the atomic weight was very nearly a whole number, was found to consist of atoms of only one atomic weight identical to that of the element. Aston was thus led to formulate his 'whole number rule' according to which the atomic weights of all atoms are very close to whole numbers, and the chemically determined atomic weight of an element is equal to that of its atom or to the mean atomic weight of its several atoms.

Following Soddy, atoms of the same element but of different atomic weight are now known as 'isotopes' of the element. The whole number nearest to the atomic weight of an isotope is called its 'mass number', A. An isotope of an element may then be conveniently denoted by the chemical symbol of the element with the mass number as a superscript. More recently, oxygen has been found to have isotopic forms, although, as naturally occurring, it consists almost entirely of the isotope ^{16}O. In dealing with an isotope, therefore, it is now customary to replace atomic weight $\mathscr{A}$ by 'isotopic weight' $\mathscr{M}$, which is the weight of an atom of the isotope relative to that of an atom of ^{12}C as 12.00000. The mean isotopic weight of naturally occurring oxygen is then 15.9994 so that atomic weights on the natural oxygen scale are converted to the ^{12}C one on multiplication by

$$15.9994/16.00000 = 0.99996$$

Isotopic weights are said to be measured in 'atomic mass units', written 'amu', so that one amu is 1/12 of the mass of an atom of carbon-12. Also, since the number of atoms per gram atom is the same as the atomic weight times the number of atoms per gram, it follows that the Avogadro Number N_0 may equally well be expressed in terms of amu per gram.

The numbers of atoms of the different isotopes per atom of the naturally occurring element are known as their 'fractional abundances', and are shown by the consistency of atomic weights to be invariable. For any one element the sum of the products of the fractional abundances and isotopic weights gives the atomic weight on the carbon-12 scale, as listed in the fourth column of the table in Appendix 2. The remaining columns give the mass numbers of the constituent isotopes, their fractional abundances and their isotopic weights.

4

Radioactive radiations

4.1 Discovery of radioactivity

With the acceptance of the electron as a universal atomic constituent that became detached in the process of ionisation, the idea that atoms were indivisible had been shown to require some qualification. At about the same time, however, a new development was taking place which, so far from qualifying this idea, was to result in its wholesale rejection. This development was the discovery made in 1896 by H. A. Becquerel. In that year, Becquerel was investigating the phenomenon of fluorescence, exhibited by certain substances on exposure to light. Quite by chance, he found that a photographic plate wrapped in black paper was as strongly affected by a uranium salt whilst kept in the darkness of a drawer as it would be if exposed to sunlight. It thus appeared that the uranium salt emitted continuously and of its own accord some sort of radiation that could penetrate opaque materials and affect a photographic plate. To this phenomenon Marie Curie gave, in 1898, the name 'radioactivity'.

Immediately following discovery of the radiation itself, Becquerel established a further property of it, namely its ability to induce electrical conductivity in gases. For this purpose, he used a 'gold leaf electroscope', as shown diagrammatically in Figure 4.1. In its simplest form, this instrument consists of a vertical metal rod R integral with a metal plate P at its upper end. To its lower end are attached two small strips of gold leaf L. The rod is supported by means of an insulating bush I in an earthed metal box B, which protects the gold leaves from air currents. When an electric charge is applied to the plate P, it is conducted by the rod R to the gold leaves L. Since the leaves now carry a charge of the same sign, they repel one another to form an

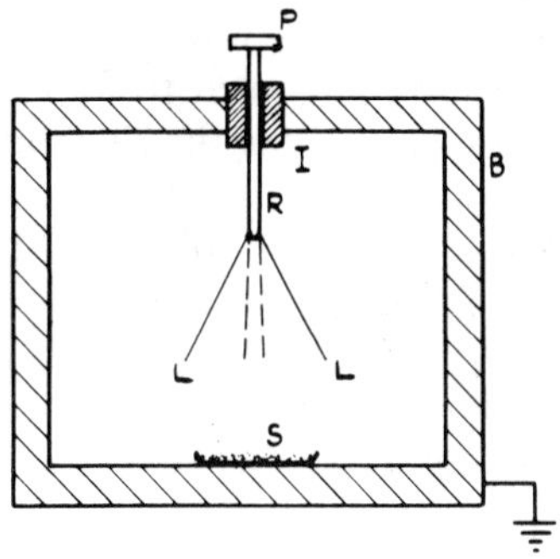

Figure 4.1 Gold leaf electroscope

inverted V as shown. Becquerel found that if a quantity of uranium salt S were placed near the leaves, they rapidly collapsed into the position shown by the dotted lines, thus indicating that the air surrounding them had been made electrically conducting so that the charge on them could leak away.

4.2 **Ionisation of gases**

The production of electrical conductivity in gases was studied in the years 1897 to 1899 by Ernest, afterwards Lord, Rutherford and J. S. Townsend. They put forward the theory that radioactive radiation creates equal numbers of positive and negative ions in a gas, analogous to the ions responsible for electrical conductivity in liquids.

The relationship between the current transmitted by an ionised gas and the applied voltage may be studied by means of an apparatus as outlined in Figure 4.2. A gas, such as air, is ionised by a radioactive

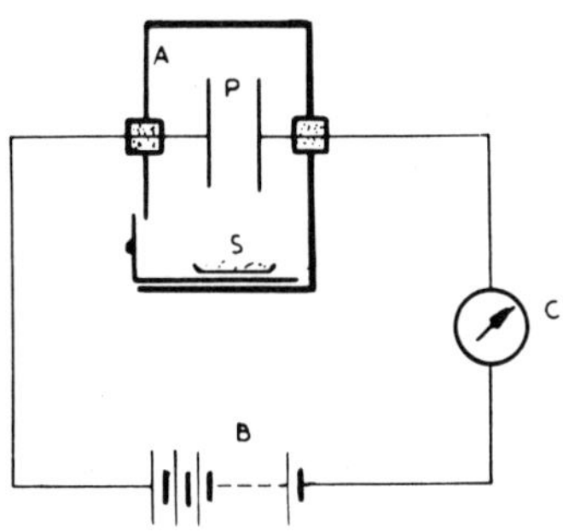

Figure 4.2 Apparatus for measuring ionisation current

source S in a metal chamber A. Two parallel metal plates P are arranged in the chamber and insulated from it, to act as electrodes. The electrodes are connected outside the chamber to a battery B and an instrument C for measuring electric current. The current passing between the electrodes is then found to vary with the applied voltage after the manner shown in Figure 4.3. It might at first be thought that the negative and positive ions would reach their respective electrodes at a rate equal to that at which the air was ionised. If, therefore, the rate of ionisation were constant, the current between the electrodes would be expected to have a constant value, independent of the applied voltage. As may be seen from Figure 4.3, the current does in fact behave very nearly in this way over a wide range of applied voltage, differing only slightly from the constant value indicated by the dotted line, known as the 'saturation current'.

At low values of applied voltage, however, the current increases rapidly with voltage. This behaviour can be attributed to a tendency of gaseous ion pairs to recombine. At low applied voltages the ions move relatively slowly to the electrodes and thus have considerable opportunity for recombination. The rate at which ions reach the electrodes would thus be less than the rate of ionisation and the current between the electrodes correspondingly less than the saturation value. With increasing voltage, however, the ions are driven to the electrodes more and more rapidly so that recombination would take place on an ever diminishing scale. The rate at which ions reach the electrodes would then approach the rate of ionisation and the current between the electrodes would tend to the saturation level.

That gaseous ions do in fact recombine was demonstrated by

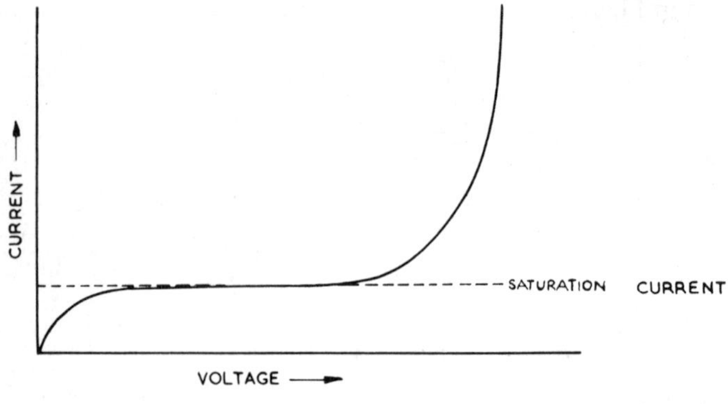

Figure 4.3 Relation between ionisation current and applied voltage

Rutherford by means of the apparatus shown diagrammatically in Figure 4.4. A radioactive source S in a metal tube T was imagined to produce ion pairs in the air above it at a rate q per unit volume. The air was drawn through the tube which was sufficiently wide to reduce losses of ions to it to negligible proportions. A cotton plug P filtered out any ions in the air at entry to the tube by causing them to make contact with the tube walls and there give up their charge. If there were n ion pairs per unit volume of air then, in this volume, the rate of collision of positive ions with a given negative ion would be proportional to n, so that the rate of collision per unit volume between positive and negative ions would be proportional to n^2, or equal to $\alpha_r n^2$, where α_r was called the 'coefficient of recombination'. It followed that the rate of increase of ions per unit volume at time t would be

$$dn/dt = q - \alpha_r n^2$$

At time t after the unit volume of air had moved out of the influence of the radioactive source,

$$dn/dt = -\alpha_r n^2$$

whence

$$n = n_0/(1 + n_0 \alpha_r t)$$

where n_0 denoted the initial value of n. By connecting each of the electrodes E_1, E_2, E_3 in turn to the battery B, the corresponding

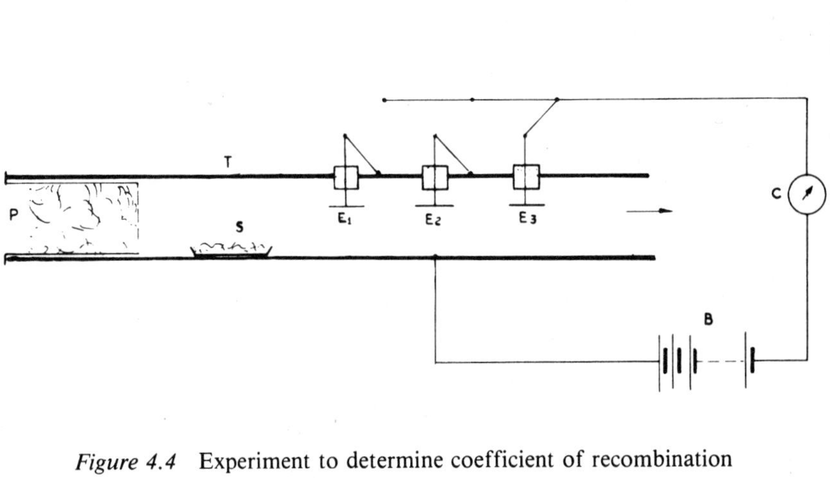

Figure 4.4 Experiment to determine coefficient of recombination

saturation currents were measured by the instrument C. These currents, together with the assumption that each ion carried a single electronic charge, gave directly the rates at which ions reached the electrodes, and these rates were divided by the air flow to give the appropriate values of n. Taking the value of n at electrode E_1 as n_0 and the corresponding value of t as 0, the values of t appropriate to the other electrodes were obtained from their distance apart and the air velocity. The value of α_r was then readily derived from the above equation for n.

At a sufficiently high applied voltage, it is seen from Figure 4.3 that the current transmitted by an ionised gas again rapidly increases. This increase was first studied by J. S. Townsend by means of the apparatus outlined in Figure 4.5. The vessel V contained gas at a low pressure. Ultra-violet light entered the vessel by the quartz window W and passed through the electrode A. This electrode consisted of a thin slab of quartz covered with a layer of silver with transparent scratches upon it to allow the ultra-violet light to pass through. After passing through the electrode A, the ultra-violet light fell on a second electrode C, consisting of a flat zinc plate, causing photo-electrons to be emitted. Since the electrode C was maintained at a negative potential, the photo-electrons were repelled from it and were assumed to attach themselves at once to molecules, forming negative molecular ions at a rate m_0. This rate of formation of ions was taken to be analogous to that due to the presence of a radioactive source, the absence of the possibility of recombination being unimportant at the applied voltages of interest. Townsend assumed that each negative ion made successive collisions to produce α new ion pairs per unit perpen-

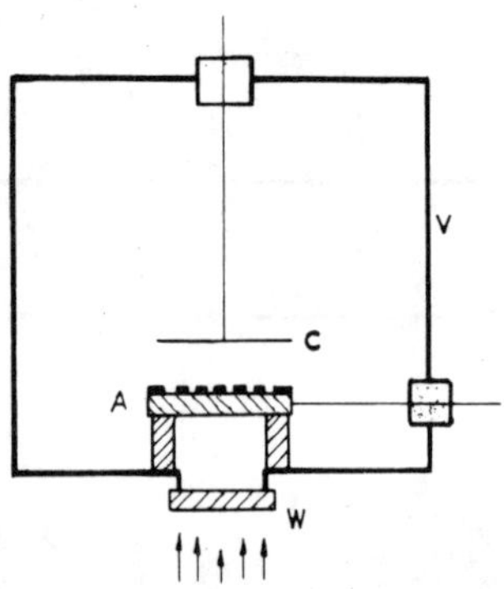

Figure 4.5 Apparatus for studying ionisation by collision

dicular distance between the electrodes. In the same distance, each positive ion likewise produced β new ion pairs. Referring to Figure 4.6, p denoted the rate at which ion pairs were generated by collision over a distance x from the electrode C. The rate at which negative ions passed through a plane at x parallel to the electrodes was thus $m_0 + p$. The corresponding rate at which positive ions passed in the opposite direction was $p_l - p$, p_l being the value of p at a distance x equal to l, the gap between the electrodes. The rate at which ion pairs were created over the distance dx was then

$$\mathrm{d}p = (m_0 + p)\,\alpha \mathrm{d}x + (p_l - p)\,\beta \mathrm{d}x$$

Since p is 0 when x is, the solution to this equation is

$$p = \frac{m_0\alpha + p_l\beta}{\alpha - \beta}\,[\mathrm{e}^{(\alpha - \beta)x} -]$$

whence

$$m_0 + p_l = m_0 \frac{(\alpha - \beta)\,\mathrm{e}^{(\alpha - \beta)l}}{\alpha - \beta\,\mathrm{e}^{(\alpha - \beta)l}}$$

and is the rate at which negative ions would reach the electrode A.

On the assumption that each ion carried one electronic charge, the current between the electrodes at a low applied voltage gave the value of m_0. At voltages sufficiently high for ionisation by collision, the current between the electrodes gave the value of $m_0 + p_l$. By varying the gap l, and adjusting the voltage to maintain the electric field between the electrodes and therewith α and β unchanged, the quantity $m_0 + p_l$ was found to vary in accordance with the formula derived above. Although more recent investigations have thrown doubt on Townsend's β mechanism of positive ion collision, his work shows that the increase in current through an ionised gas at very high applied

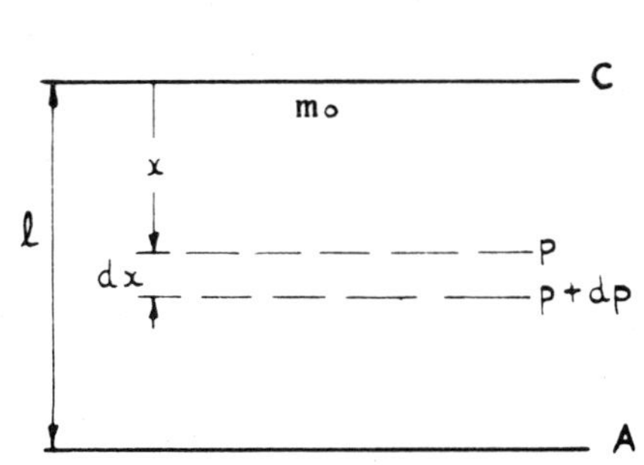

Figure 4.6 Diagram for theory of ionisation by collision

voltages is attributable to the ions acquiring sufficient energy between collisions to enable them to ionise a gas molecule when collision occurs. On collision, the ion ejects an electron from the gas molecule, thereby creating an ion pair. These newly formed ions are rapidly accelerated by the electric field and themselves ionise further gas molecules by collision so that the rate of ionisation of the gas is greatly increased.

4.3 Types of radiation

The ability of radioactive radiations to ionise, together with the insight obtained into the ionisation process itself, made possible comparative measurements of their intensity. At the same time, however, other properties of these radiations were being discovered. In 1899, Rutherford studied the extent to which the ionising power of the radiations was affected by causing them to pass through various thicknesses of aluminium. He found that the ionising power of a uranium compound was markedly reduced by a thickness of only 0.03 mm or so but that thereafter it decreased much more gradually, vanishing only when the thickness had reached about 3 mm. From these results Rutherford concluded that the radiations were of two types, one of very small penetrating power that produced the greater part of the ionisation and which he called 'alpha (α) rays', and the other of much greater penetrating power but producing much less ionisation which he called 'beta (β) rays'. This conclusion received support in 1899 when Becquerel, simultaneously with S. Meyer, E. von Schweidler and F. Giesel, found that part of the radioactive radiations was readily deflected in a magnetic field whilst the remaining part appeared not to be. By comparing the penetrating powers of these two parts, Marie Curie showed that the deflected part was identical with Rutherford's β-rays. The magnetic deflection of the β-rays was such as to suggest that they were associated with a negative electric charge, and this association was confirmed a year later when Becquerel made approximate measurements of their speed and charge-to-mass ratio from their deflections in electric and magnetic fields. He thereby concluded that the β-rays were identical with cathode rays.

In the same year, however, a more accurate experiment was made by W. Kaufmann with the apparatus shown schematically in Figure 4.7. A small piece of radioactive material S was placed in a highly evacuated vessel V. The radiations from this material passed between electrodes in the form of parallel plates P, and then through a hole in the diaphragm H to strike a photographic plate D. The whole

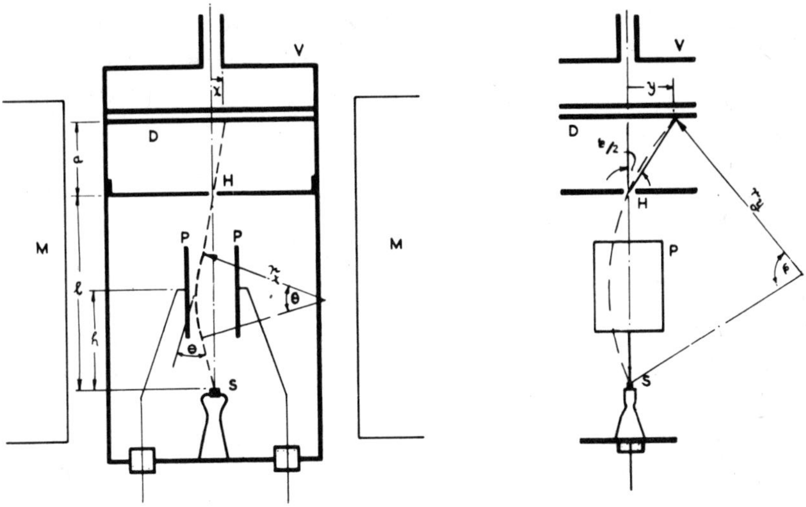

Figure 4.7 Apparatus to determine charge-to-mass ratio of β-particle

apparatus was maintained in a uniform magnetic field imposed by the magnet MM. At the photographic plate, the β-rays experienced a deflection x in a direction normal to the parallel plates due to the electric field between them, and a deflection y in a plane parallel to the plates due to the magnetic field. If the electric field were of strength X, then a β-particle having mass m, charge e and velocity v would be deflected into a circular path of radius r_x, such that

$$Xe = mv^2/r_x$$

by equating electric and centrifugal forces. Similarly, if the magnetic field were of flux density B, the β-particle would be deflected into a circular path of radius r_y, such that

$$Bev = mv^2/r_y$$

by equating magnetic and centrifugal forces.

If the circular path of radius r_x subtended an angle θ, and if the centre of plates P and the hole H were distant h and l respectively above the radioactive material S, then the deflection at the photographic plate, distant d above the hole H, would be

$$x = \frac{\mathrm{d}}{l} h\theta$$

$$\propto 1/r_x$$

Similarly, if the circular path of radius r_y subtended an angle φ, then the corresponding deflection at the photographic plate would be

$$y = \varphi d/2$$
$$\propto 1/r_y$$

Hence,

$$y^2/x \propto (B^2/X) \, . \, (e/m)$$

the constant of proportionality depending only on the geometrical arrangement of the apparatus. As in Thomson's parabola method described in Section 3.2, therefore, all particles having the same ratio of charge to mass, whatever their velocity, would form spots on the photographic plate tracing a unique parabolic curve. By measuring the coordinates for any point on the curve, the charge-to-mass ratio could be determined. In this way Kaufmann established that the charge-to-mass ratio of the β-particle was the same as that of the electron, thereby confirming the identity of these two particles.

Although early experiments led to the belief that α-rays could not be deflected in a magnetic field, the stronger fields used by Rutherford in 1903 revealed that they could be deflected in this way. The deflection was such as to indicate that α-rays consist of positively charged particles.

In 1900, P. Villard discovered that radioactive substances were capable of producing a weak ionisation of a gas even when separated from it by several centimetres of iron. The radiations responsible for this effect were clearly far too penetrating to be identified with α-rays. Furthermore, they differed from β-rays as evidenced by failure to deflect them with the strongest available magnetic field. These radiations thus constitute a third distinct type and are called 'gamma (γ) rays'. Like α- and β-rays, γ-rays are capable of affecting a photographic plate.

The different electrical properties of the three types of radioactive radiation may be summarised as in Figure 4.8, which reproduces a diagram published by Marie Curie in 1903. A radioactive source is supposed to be placed in a deep narrow cavity in a block of lead. In the absence of electric and magnetic fields, the radioactive radiations would then emerge as a straight narrow beam. On application of a strong magnetic field in a direction at right angles to this beam, however, the α- and β-rays would be deflected in opposite directions in a plane normal to the field, the α-rays less strongly and the β-rays more so. The γ-rays, on the other hand, would remain completely unaffected.

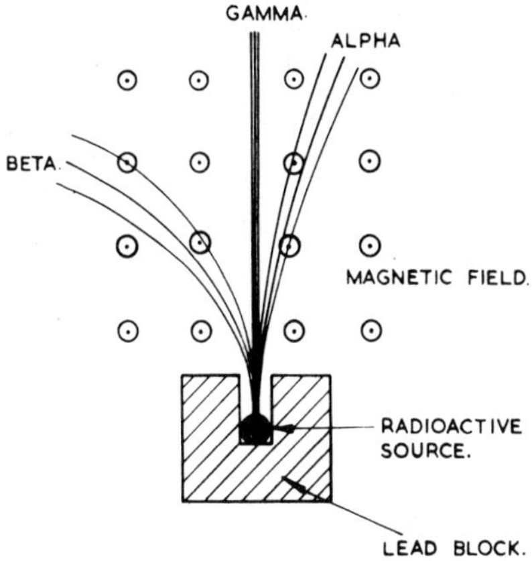

Figure 4.8 Representation of radioactive radiations

The differences in penetrating power of the three types of radiation have already been noted, the α-rays penetrating only slightly, the β-rays more readily and the γ-rays very strongly indeed. From the above account of these rays, however, it is also clear that the more penetrating they are, the less powerfully are they able to produce ionisation in gases.

4.4 The ionisation chamber

The apparatus shown diagrammatically in Figure 4.2, which was described in Section 4.2, constitutes a simple type of 'ionisation chamber'. A chamber of this type was used by Marie Curie in 1898 in some of the earliest quantitative studies of radioactivity, and is especially suited to the counting of α-particles in the presence of other radiations. Since α-particles are easily absorbed, even in air, the plates P are arranged only 1 cm or so apart, and the substance under examination placed as closely as possible to them. In this way, the intensity of the α-radiation is maintained at a high level and its superior ionising power then renders the effects of β- and γ-rays negligible by comparison. The battery voltage is adjusted so that the current is at its saturation value. The pulse of current produced by an α-particle is then unaffected by small changes in voltage. The response of such a shallow ionisation chamber is usually so rapid that α-particles reaching it at intervals as short as 10^{-4} second or less can be individually detected.

For the measurement of β- and γ-radiations, the ionisation chamber is modified so as to determine the total ionisation produced rather than the number of individual particles. Referring to Figure 4.9, the radiation enters a cylindrical metal chamber C through a window W consisting of a thin sheet of aluminium. This window prevents the entry of α-particles without materially impeding β- or γ-radiation. The chamber contains atmospheric air and forms one of the electrodes. Within the chamber and insulated from it is a thin metal rod E which forms the other electrode. The circuit outside the chamber is completed by the battery B and a current-measuring instrument. In this form, the chamber responds chiefly to β-rays. Allowance must, however, be made for any γ-rays present. In order to measure the intensity of γ-rays, the chamber is usually filled with a heavy gas, such as argon or 'Freon', often at high pressure, so as to facilitate ionisation. Any α- or β-rays present may then be cut out, with little influence on the γ-rays, by means of a thin sheet of lead.

Originally the pulses of current from an ionisation chamber were measured directly by means of an electrometer but later it was more usual to pass them through a vacuum-tube amplifier. Then, in the case of the chamber shown in Figure 4.9, the ionisation current raises the potential at the end of resistance R nearest the electrode E. It can thus be amplified by the amplifier A and transmitted to a measuring instrument or mechanical counter.

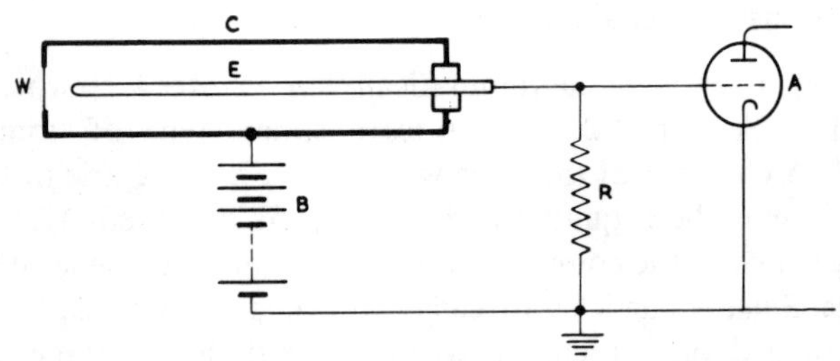

Figure 4.9 Ionisation chamber

5

Radioactive decay

5.1 Natural radioelements

As mentioned in Section 4.4, some of the earliest quantitative studies of radioactivity were made in 1898 by Marie Curie using a parallel plate type of ionisation chamber. She set out to find whether elements other than uranium were radioactive and, simultaneously with G. C. Schmidt, found that thorium was. A far-reaching discovery, however, awaited her examination of pitchblende, an oxide of uranium, for its rate of emission of radiation, or 'activity', was found to be many times greater than that from an equal weight of uranium itself. Marie Curie was thus led to believe that pitchblende contained an element much more active than uranium, and in the same year her husband Pierre Curie joined her in the search for it. By repeated chemical processing of a quantity of pitchblende they obtained more and more highly active substances, finally arriving at a hitherto unknown metal of unprecedented activity, to which they gave the name 'polonium'.

Still in the year 1898, the Curies found that pitchblende contained a second radioactive substance chemically different from polonium and of even greater activity. To this new element they gave the name 'radium'. It was a small sample of radium that Kaufmann used in identifying the β-particle with the electron. In 1902, the Curies put the existence of radium beyond all doubt by securing a ton of pitchblende residues and extracting from it, under the most primitive conditions, one tenth of a gram of radium chloride, sufficient to determine the atomic weight of radium. The chemical properties of radium were seen to be analogous to those of barium and in the same column of the periodic table radium fell naturally into place.

By now, the Curies were convinced that radioactivity was a property of an atom itself, unaffected as it was by any chemical change in which the atom took part. Furthermore, they had recognised the ionising power of the radiations from radioactive atoms as a clear indication that these atoms were constantly liberating energy. In the case of radium the energy liberated was measured in 1903 by Pierre Curie and A. Laborde, who found it to be over 100 cal/g hour, or as much in $2\frac{1}{2}$ days as is evolved by the combustion of an equal weight of coal. Moreover, this rate of energy release appeared scarcely to diminish with the passage of time, so that the immense amount of energy locked up within the atom began to be revealed. The problem of releasing this energy more quickly, however, was to remain unsolved for forty years.

In 1899, following the discovery of polonium and radium by the Curies, a new radioactive element was found in pitchblende residues by A. Debierne and given the name 'actinium'. Including uranium and thorium, therefore, there were known at this time to be five different radioactive elements or 'radioelements' as they are now called. By 1904, largely due to Ernest Rutherford and Frederick Soddy, twenty such elements had been identified and this number had increased to over thirty by 1912. Today, over forty different radioelements are known to occur in nature.

5.2 **Uranium X**

In 1900, Sir William Crookes employed several simple chemical methods to remove from a uranium compound a constituent which he called 'uranium X' and which, judged by its effect on a photographic plate, appeared to possess all the radioactivity resident in the original compound. At the same time, the compound from which uranium X had been removed was found to be very nearly inactive. Crookes was therefore led to suggest that the activity of uranium compounds was due to an impurity in them and not to the element uranium itself. Since the activity of a given uranium compound, however prepared, was invariably the same, this suggestion was unacceptable to Becquerel and, in 1901, he obtained conclusive evidence against it. Becquerel found that if uranium X and the uranium compound from which it had been removed were laid aside separately for eighteen months, then at the end of that time, the compound had completely regained its former activity whilst the uranium X appeared to be completely inactive.

These effects were studied quantitatively by Rutherford and Soddy, and their first results were in direct contrast to those obtained by Crookes. For they found that the uranium from which uranium X had been separated suffered no reduction in activity, as measured by its ionising effect. However, on replacing the ionisation method of detecting radiation by the photographic one used by Crookes, they obtained the same results as he had. It thus transpired that the photographic effect of a uranium compound is almost entirely due to β-rays and that the changes in its activity and in that of uranium X noted by Crookes and Becquerel relate only to this type of radiation. The α-radiation of uranium remains totally unaffected.

Turning to a more detailed examination of the way in which the β-activity decayed in uranium X and recovered in the uranium compound from which it had been removed, Rutherford and Soddy obtained in 1902 the results displayed graphically in Figure 5.1. These results show that the activity of uranium X decays exponentially with time whilst that of the uranium compound recovers in a complementary manner so that the simultaneous rates of decay and recovery are equal. The combined activity of the uranium X and the uranium compound from which it has been removed thus remains unchanged. In the same year, Rutherford and Soddy found that a

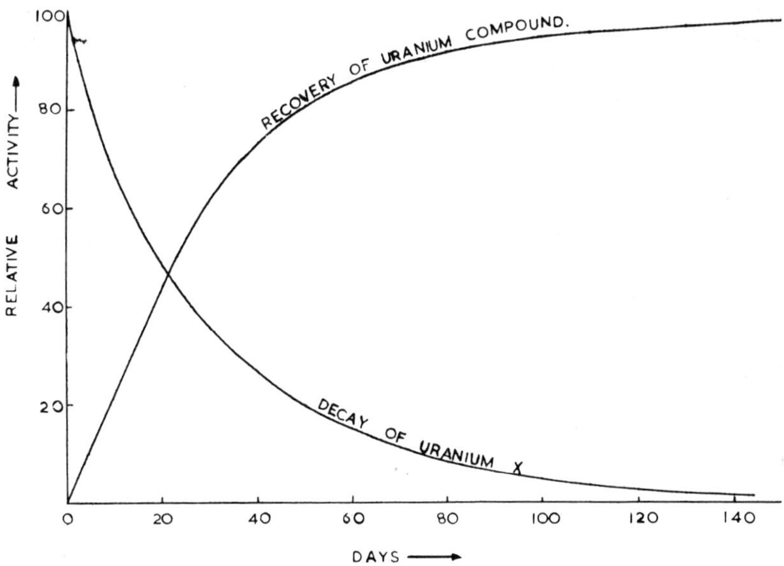

Figure 5.1 Decay of uranium X and recovery of uranium compound

thorium compound behaved similarly in relation to a constituent that they called 'thorium X' analogous to uranium X. The interpretation of these results had, however, to take account of other observations that were being made at this time.

5.3 Emanations and active deposits

In 1899, R. B. Owens noted that the radioactivity of thorium appeared to be influenced by neighbouring currents of air. A year later Rutherford showed that this influence was due to the emission by thorium compounds not only of radioactive radiation but also of a radioactive 'emanation' or vapour, which emitted only α-rays. The apparatus used by Rutherford to study this emanation is illustrated diagrammatically in Figure 5.2. The thorium compound S was enclosed in a tube A through which dried air flowed to a plug of glass wool in a bulb P. This plug removed any ions originally present in the air, which then passed into the testing cylinder, T, to flow successively past electrodes E_1, E_2, E_3. The electrodes were insulated from the testing cylinder and connected to it externally through a battery B, either directly or through a current-measuring instrument C. The battery voltage was such that the ionisation current at each electrode reached its saturation value, and was thus directly proportional to the corresponding activity of the emanation. This activity was found to decay in an exponential manner.

It was further noted by Rutherford that all surfaces in contact with the emanation eventually developed an 'active deposit', emitting all three types of radiation. This deposit was removable either by sand-papering or by dissolving in acids and its activity, although it decayed, did so more slowly than that of the emanation. Marie and Pierre Curie

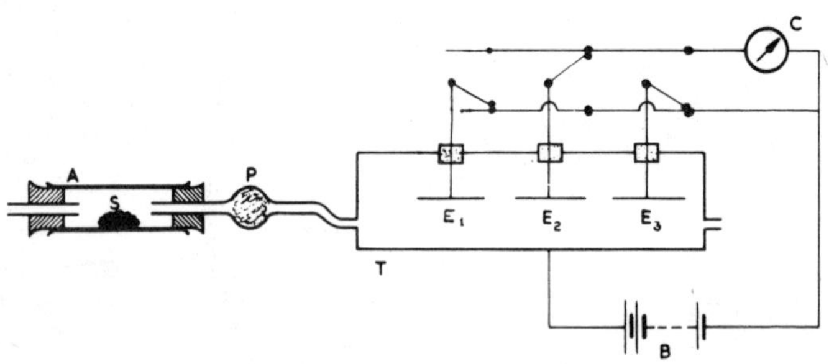

Figure 5.2 Apparatus for studying thorium emanation

had also reported, in 1899, that radium compounds gave rise to an active deposit on nearby surfaces, and a year later E. Dorn had proved that these compounds gave off an emanation as well. A corresponding actinium emanation was discovered in 1903 by Debierne. Rutherford and Soddy found these emanations to behave like ordinary gases and succeeded in condensing them by liquid air at about $-150°C$.

5.4 Laws of radioactive decay

The behaviour of uranium X and thorium X, and the formation of the emanations and active deposits, were all correlated in 1903 when Rutherford and Soddy put forward their theory of 'radioactive disintegration'. According to this theory the atom of a radioelement undergoes spontaneous disintegration, with the emission of an α- or β-particle, to become an atom of a new element. This atom may also be radioactive and undergo successive disintegrations until a non-radioactive, or 'stable' element is reached. At the time at which this theory was proposed, the three radioactive series of uranium, radium and thorium appeared to be as in Figure 5.3. The direction of the change from one element to another is indicated by the arrows and the type of particle emitted in the disintegration is shown to one side of them. The first element of a radioactive series is known as the 'parent', the second the 'first daughter product', the third the 'second daughter product' and so on.

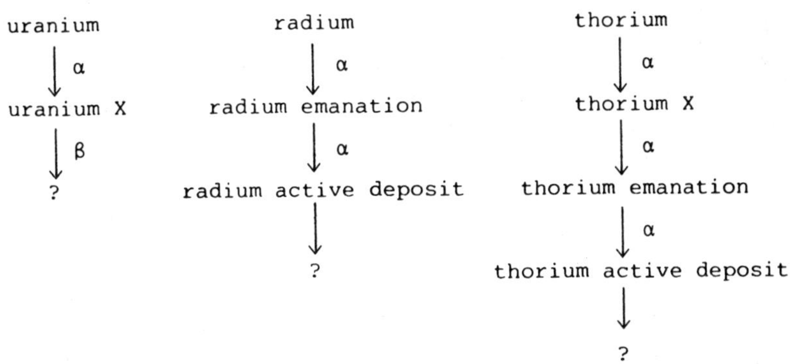

Figure 5.3

From their results as shown in Figure 5.1, it appeared to Rutherford and Soddy that the activity of a radioelement, separated from its source of formation, decreased exponentially with time. Now

$$\text{activity} = \text{rate of disintegration} = -\mathrm{d}P/\mathrm{d}t$$

where P is the number of atoms of the radioelement at time t. It followed, therefore, that

$$-\mathrm{d}P/\mathrm{d}t \propto \mathrm{e}^{-\lambda t}$$

where λ is a constant. Integrating,

$$P = P'\mathrm{e}^{-\lambda t} \tag{3}$$

where P' is the initial value of P. Also, the activity

$$-\mathrm{d}P/\mathrm{d}t = \lambda P \tag{4}$$

and is directly proportional to the number of atoms of the radioelement present. Activity is commonly measured in terms of the 'curie', equal to 3.7×10^{10} disintegrations per second.

The constant λ is known as the 'decay constant', and has a value characteristic for each radioelement unaffected by any temperature, pressure, phase, or chemical changes in which it may take part. It follows from equation (3) that the number of atoms of a radio-element, separated from its source of formation, will be reduced by a factor 2 in a time interval

$$t_{\frac{1}{2}} = 0.693/\lambda \tag{5}$$

The interval $t_{\frac{1}{2}}$ is known as the 'half-life' and was introduced by Rutherford in 1904 as an alternative to the decay constant.

The recovery of a uranium compound from which uranium X has been removed is also governed by equation (4). For if suffixes u and x denote uranium and uranium X respectively, then at time t after removal of the uranium X,

$$\left.\begin{aligned} \mathrm{d}P_u/\mathrm{d}t &= -\lambda_u P_u \\ \mathrm{d}P_x/\mathrm{d}t &= \lambda_u P_u - \lambda_x P_x \end{aligned}\right\}$$

whence

$$\begin{aligned} P_x &= \frac{\lambda_u}{\lambda_x - \lambda_u} P_u' \left(\mathrm{e}^{-\lambda_u t} - \mathrm{e}^{-\lambda_x t}\right) \\ &\approx \frac{\lambda_u}{\lambda_x} P_u' \left(1 - \mathrm{e}^{-\lambda_x t}\right) \end{aligned}$$

if λ_u is very small, as is the case. This relationship is seen to be of the same form as that described by the recovery curve in Figure 5.1.

The half-life of a radioelement may be determined by measuring the relative activity at different instants of time. By equations (3) and (4),

$$\text{activity} \propto e^{-\lambda t}$$

so that a plot of log (activity) against time t results in a straight line of negative slope λ. Measurement of this slope then gives the required half-life. If two or more radioelements are present, the slope will vary with time, but by analysing the variation it may be possible to determine the corresponding half-lives. This method is not practicable if the half-life is very long. To obtain an alternative method the analysis given above of the recovery of a uranium compound is generalised so that

$$\left.\begin{aligned} dP_0/dt &= -\lambda_0 P_0 \\ dP_1/dt &= \lambda_0 P_0 - \lambda_1 P_1 \\ dP_2/dt &= \lambda_1 P_1 - \lambda_2 P_2 \\ &\dots \\ dP_s/dt &= \lambda_{s-1} P_{s-1} \end{aligned}\right\} \qquad (6)$$

where suffix 0 denotes the parent element, suffix 1 the first daughter product, suffix 2 the second daughter product and suffix s the stable end product. If the half-life of the parent is very long in comparison with those of succeeding elements of the series, then these elements will behave as though the activity $\lambda_0 P_0$ were constant. After a sufficient length of time, therefore, they will reach an equilibrium concentration such that

$$\lambda_0 P_0 = \lambda_1 P_1 = \lambda_2 P_2 = \dots$$

and are then said to be in 'radioactive equilibrium'. If the half-life of only one of these elements is known, then their relative atomic concentrations at equilibrium give the half-lives of the remainder.

6

The alpha particle

6.1 The Geiger counter

Establishment of the β-particle as an electron shed little new light on the structure of the atom since Thomson had already concluded that the electron was a separable constituent of all atoms. The discovery of the α-particle raised fresh issues, however, for it was known to be positively charged and not therefore identifiable with the electron. A great deal of research, mainly by Rutherford, thus became directed towards a fuller understanding of the nature of the α-particle. In the course of this research the need arose for a more positive method of recording individual α-particles than that provided by the ionisation chamber. This need was filled in 1908 when Rutherford and H. Geiger first employed the principle that underlines the modern Geiger counter.

The apparatus used by Rutherford and Geiger is outlined in Figure 6.1. The α-particles enter by the open end of a tube T, which is connected to a vessel V. The rate at which α-particles enter the vessel is regulated by means of a plug valve P. A very thin sheet of mica over a perforated diaphragm H permits adjustment of the pressure of the gas

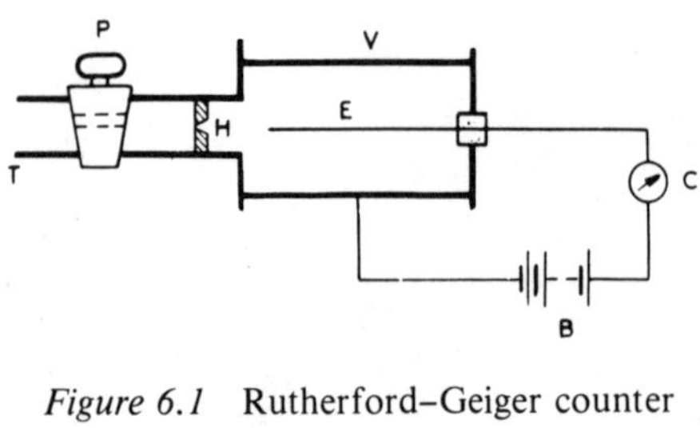

Figure 6.1 Rutherford–Geiger counter

in the vessel V to a few centimetres of mercury. Along the axis of the vessel, and insulated from it, is a wire E which forms one of the electrodes. The other electrode is provided by the walls of the vessel itself. Outside the vessel, the circuit is completed by a battery B and an electrometer C. The gas pressure is so adjusted that the number of ions created by each α-particle is greatly multiplied through ionisation by collision. The pulse of current thereby produced is then readily detected.

Geiger improved this type of counter in 1913, but in 1928 he collaborated with W. Müller to produce the modern form which is known as a 'Geiger–Müller' or 'G–M' counter. It is highly sensitive and will count β-particles as well as α-particles. In this counter ionisation by collision occurs on such a scale that the resulting pulse of current is almost independent of the number of ions directly created by the entering ionising particle. The counter is thus unable to distinguish between one type of ionising particle and another. It can be used to count β-particles in the presence of α-particles, however, if the radiation is made to enter it through an aluminium window, which effectively stops the α-particles but lets the β-particles through.

6.2 Scintillation counting

The study now made of the interaction between α-particles and matter achieved its most significant results not with the Geiger counter but with the 'scintillation' method of counting, which had the advantage of comparative insensitivity to β- and γ-radiation. This method arose from a discovery made by Becquerel in 1899 that radioactive radiations were able to produce luminescence in certain substances, notably zinc sulphide. Four years later, Crookes found that the luminescence produced by α-particles on zinc sulphide was not uniform but consisted of a large number of discrete flashes. Each flash was attributed to the impact of a single α-particle and thus provided the earliest means of individual α-particle counting. In making this discovery, Crookes used an instrument which he called a 'spinthariscope', having the form shown diagrammatically in Figure 6.2. A cylindrical casing C contained a zinc sulphide screen D at one end and was fitted with an eyepiece lens E at the other. The screen was bombarded by α-particles emitted by a trace of radium at the point of a needle S. The height of the needle above the screen could be adjusted by a screw W. With the eyepiece lens correctly focused, the luminescent glow on the screen was resolved into a myriad of discrete scintillations.

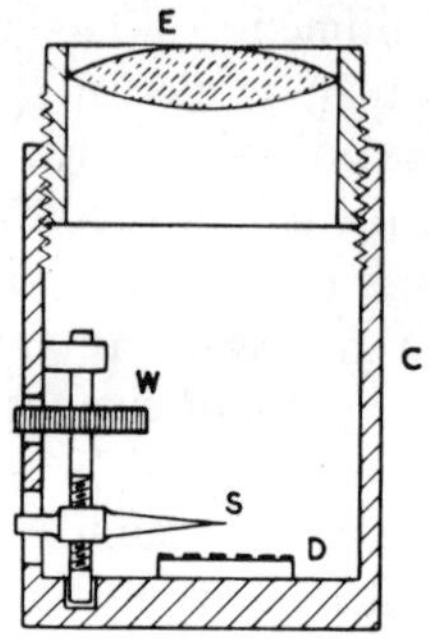

Figure 6.2 The spinthariscope

In 1908, Rutherford and Geiger compared the number of scintillations produced by a radium source on a zinc sulphide screen with the number of current pulses produced by the same source in a Rutherford–Geiger counter. The numbers were approximately the same and the reliability of the scintillation method of counting α-particles was thereby established. The scintillation method, relying on tedious visual observation, was almost exclusively used up to the beginning of the third decade of this century. Thereafter, with the development of vacuum-tube circuits, the scintillation method gave place to methods of counting α-particles by means of their associated ionisation current pulses. More recently, however, interest in the scintillation method has revived due to the advent of the electron multiplier tube with a photo-electric cathode, which enables the feeble light of a scintillation to produce a measurable pulse of current.

6.3 Arrest of α-particles by matter

In 1904, Sir William H. Bragg examined the ionisation produced in air by α-rays at various distances from their source. The apparatus used is outlined in Figure 6.3. The α-rays from a source S emerged by the small aperture H in a lead box so as to form a narrow divergent beam. Two metallic gauzes P were arranged close and parallel to one another and formed an ionisation chamber. The saturation current between the gauzes was measured, thereby giving the rate of ionisation per unit length of α-ray path. By varying the distance of the ionisation chamber from the source, this rate was measured at every point in the α-ray path. At a sufficient distance from the source the rate of ionisation dropped rapidly to zero. This distance therefore represented the

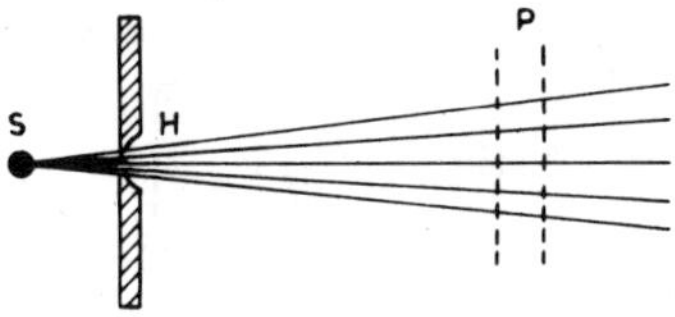

Figure 6.3 Apparatus for determining range of α-particles

limit of travel of the α-particles in air. The maximum distance over which a particle can travel in atmospheric air beyond its existing position is termed its 'range'.

A year later, Bragg and R. D. Kleeman investigated the reduction in range of α-particles on travelling various distances through solids and gases other than air. The reduction in range of an α-particle due to a material in its path is referred to as the 'stopping power' of the material. Alternatively, the 'mass stopping power' is the mass of the material per unit area normal to the direction of the α-rays divided into the corresponding mass of atmospheric air that would produce the same reduction in range. To a first approximation, the mass stopping power of a material was found to be inversely proportional to the square root of its atomic weight, a result known as Bragg and Kleeman's Law. The stopping power of a material depends to a small extent on the range of the α-particles just before entering and just after leaving the material, but when these ranges are respectively the maximum range and zero, then the mass stopping power is approximately

$$\begin{aligned} S &= \varrho_0 R_0 / \varrho T \\ &= 3.8/\sqrt{\mathscr{A}} \end{aligned}$$

where ϱ, T and $\mathscr{A}$ denote respectively the density, thickness and atomic weight of the material in question, ϱ_0 the denstity of atmospheric air and R_0 the maximum range of the α-particles.

6.4 Mass, charge and velocity of the α-particle

In 1906, Rutherford examined the relationship between the range of an α-particle and its velocity by means of the apparatus shown diagrammatically in Figure 6.4. The source of α-particles was radium active deposit on a wire S that had previously been exposed to radium emanation. The wire was mounted in an evacuated vessel V to lie parallel with a slit H, beyond which was a photographic plate D. A

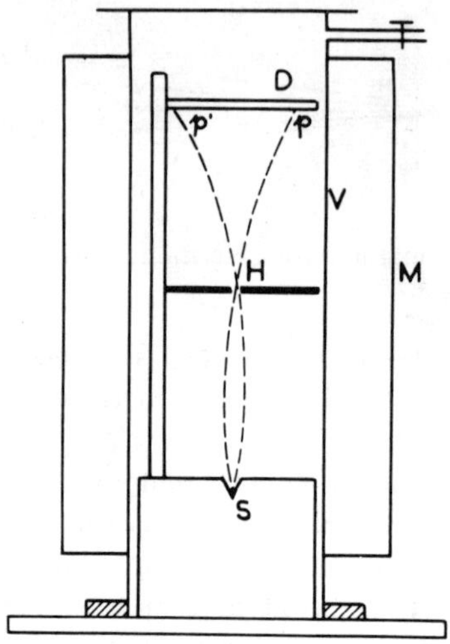

Figure 6.4 Apparatus for magnetic deflection of α-particles

magnet M subjected the whole of the vessel to a strong uniform magnetic field in the direction of the wire. If this field were of flux density B, then an α-particle of mass M, charge E and velocity v would be deflected into a circular path of radius r_y, such that

$$BEv = Mv^2/r_y$$

by equating magnetic and centrifugal forces. The radius r_y is determined by the dimensions of the apparatus and the line p produced by the α-particles on the photographic plate. The position of this line was established more precisely by reversing the field to obtain the corresponding line p′, and then the value of the radius was refined by calculation.

The region between the source and the photographic plate was equally divided by a vertical mica sheet at right angles to the direction of the magnetic field. Half of the wire S was then left bare and the other half covered with layers of aluminium foil of known stopping power. Two sets of markings p, p′ were thus recorded on the photographic plate, one set appropriate to the bare half of the wire and the other to the covered half. The ratio of the corresponding radii r_y then gave the velocity of the α-particles that had passed through the aluminium in terms of the velocity of those that had not. On repeating the experiment for successive layers of aluminium foil, a relation was obtained between the range of an α-particle and its velocity relative to that at maximum range.

For α-particles of any particular range, this experiment gave the ratio E/Mv. In order to obtain both E/M and v, a separate experiment was required in which the α-particle was deflected in an electric field instead of a magnetic one. The apparatus used by Rutherford for this deflection is outlined in Figure 6.5. As in the previous experiment, the α-particle source was radium active deposit on a wire S. The α-particles entered an evacuated vessel V through a thin mica plate in its base and then passed between two parallel insulated plates PP to fall on a photographic plate D. On applying an electric field of strength X between the plates PP, the outermost record on the photographic plate would be due to α-particles that grazed the top edge of the plate P nearer this record and met the other plate P tangentially, as shown for two extreme cases by the dotted lines 1–1, 2–2 in Figure 6.5. The outermost record on the photographic plate and the dimensions of the apparatus thus determined the radius r_x of the path of the α-particles in the electric field, provided that r_x did not exceed the limit indicated by line 2–2 in the figure. Reversal of the electric field enabled the

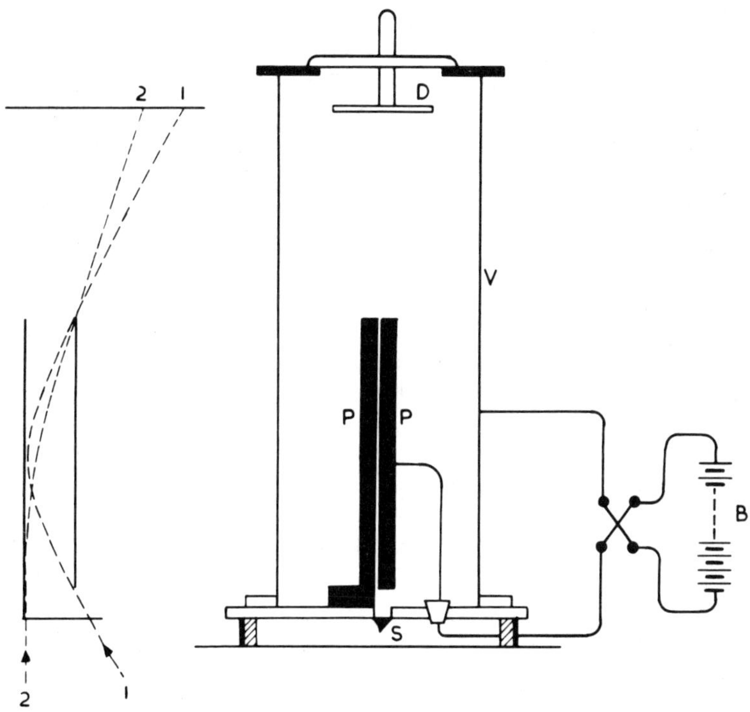

Figure 6.5 Apparatus for electric deflection of α-particles

position of the outermost record on the photographic plate to be determined with increased accuracy.

Equating electric and centrifugal forces

$$XE = Mv^2/r_x$$

The velocity v was less than the corresponding velocity for the bare wire in the magnetic deflection experiment due to the mica plate. The ratio of the two velocities was known, however, from the magnetic deflections obtained with aluminium foils and a correction for the mica plate could therefore be made by increasing X in the square of this ratio. With this correction, the last two equations gave

$$E/M = r_x X/B^2 r_y^2$$

Since the field strength X and flux density B were known, and the radii r_x and r_y were measured, the ratio E/M could be determined. It was found to be 0.84×10^{-19} coulombs/amu, whence either of the two equations containing v gave the particle velocity. This velocity was found to be very high, of the order of 7 per cent of the velocity of light.

The mass M of the α-particle could be found by measuring the charge E, and this measurement was made by Rutherford and Geiger in 1908. The apparatus used for this measurement is shown diagrammatically in Figure 6.6. A small quantity of radium was placed in a shallow dish and covered with a very thin aluminium foil to provide a radioactive source, S. The rate at which α-particles were emitted by the source was determined by the Rutherford–Geiger counter described in Section 6.1. The source was placed in a vessel V, which was evacuated to avoid collisions between the α-particles and any gaseous atoms present. The α-particles from the source fell upon the collecting plate, E, the target area on this plate being defined by the annular diaphragm H. The collecting plate was insulated from the

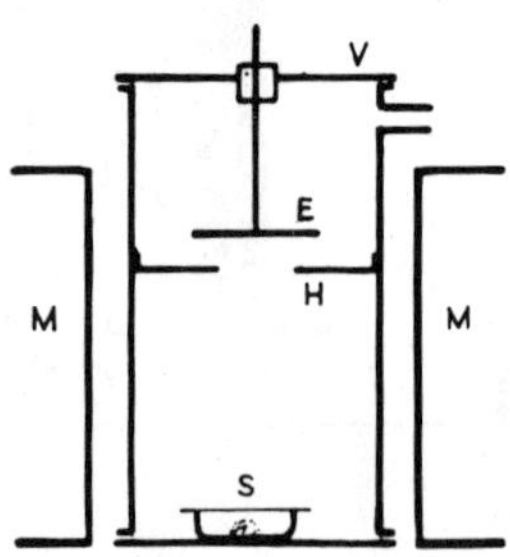

Figure 6.6 Apparatus to determine charge on α-particles

vessel V and connected outside the vessel to an electrometer. The whole apparatus was subjected to a strong magnetic field in a direction parallel to the plane of the diaphragm by a powerful magnet M. This field had little effect on the α-particles but diverted any β-particles so that they did not reach the collecting plate. The charge measured by the electrometer after a certain interval of time was noted. This charge was then divided by the number of α-particles reaching the collecting plate during the same interval, as determined initially by the counter, to give the charge carried by each α-particle. The value found for this charge was 3.1×10^{-19} coulombs, which is very nearly twice the electronic charge. The mass of the α-particle is therefore

$$\frac{3.1 \times 10^{-19}}{0.84 \times 10^{-19}} = 3.7 \text{ amu}$$

Within the limits of experimental error, this mass is that of the helium atom. It follows that the α-particle is, as Rutherford concluded, a helium atom that has lost two electrons.

With the charge-to-mass ratio of the α-particle now known, the relation obtained by Rutherford between range and velocity relative to that at maximum range immediately became one between range and absolute velocity. The magnetic deflection experiments made by Rutherford were repeated by Geiger in 1910, using mica instead of aluminium and the scintillation method of detection rather than the photographic one. Known as Geiger's Law, the relationship obtained was

$$v^3 = aR$$

where v is the particle velocity in cm/sec, R the range in cm and a has the value 1.03×10^{27} cm^2/sec^3. The velocities given by this law are accurate to within 1 per cent down to a range of 25 mm and about 10 per cent high for a range of 10 mm. At lower ranges the overestimate of velocity becomes progressively worse. However, the maximum range of α-particles from a radioactive source may be as high as 100 mm so that Geiger's Law is seen to apply over a comparatively wide variation in range.

6.5 Energy required to ionise a gas

Study of the passage of α-rays through gases led to a determination of the energy expended by an ionising particle in producing an ion pair in a gas. This determination was important since it enabled the energy of such a particle to be estimated from the current pulse produced by its arrest in the gas of an ionisation chamber.

If an ionisation chamber is so proportioned as to accommodate the entire path of a beam of α-particles entering it, then the resulting ionisation current gives the total number of ion pairs created by the beam in unit time. The rate at which the α-particles are projected along the beam can be determined by means of a Geiger or scintillation counter and the number of ion pairs created by each α-particle thus determined. The first determination of this sort was made by Geiger in 1909 from the ionisation of hydrogen by α-particles from radium active deposit.

The number of ion pairs created by an α-particle may be divided into the energy of the particle to give the energy expended by it in creating one ion pair. The energy of ionising particles is generally measured in terms of the 'electron-volt', written eV, which is the energy acquired by an electron in moving through a potential difference of 1 volt, and is equivalent to 1.6022×10^{-12} ergs. In many cases the eV unit is replaced by one of 10^6 eV, written MeV. Subsequent measurements have largely confirmed that made by Geiger and give the average energy expended by an α-particle in creating one ion pair in air as 35 eV.

The effect of the energy of the α-particle on the ionisation it produces was examined by H. Ziegert in 1928. A very sensitive electrometer was used to detect individual α-particles from various sources. In this way the average energy expended by an α-particle in creating one ion pair in air was found to vary with particle energy by only a few per cent from 35 eV. A year later, Eisl reported a value of 32.2 eV for the energy expended by electrons in creating one ion pair in air, finding no variation with electron energy over the range 10 000 to 60 000 eV.

In order to find the energy expended in creating an ion pair in a gas other than air, it is only necessary to divide the value for air by the ionisation in the gas considered relative to that in air. Relative ionisations were determined by R. W. Gurney in 1925. A beam of α-particles from a thin film of polonium passed through a mica window into an ionisation chamber which was connected to an electroscope. The range of the particles entering the chamber was varied by adjusting the pressure of gas between the source and the mica window. For residual ranges exceeding 25 mm the relative ionisations in different gases were practically independent of range and varied from 0.97 to 1.31. At smaller ranges, however, the relative ionisations in some of the gases markedly increased.

7

The structure of the atom

7.1 Rutherford's nuclear atom

With a momentum and energy over a thousand times as great as that of the electron, the α-particle was revealed as a very formidable projectile. Indeed, its effect on atoms in its path might well be expected to augment the light already shed on atomic structure by the fact of its emission. As events turned out, this expectation was more than fulfilled.

The first indication of the interaction between α-particles and atoms in their path was provided by Rutherford in 1906. He noticed that the photographic image of α-particles that had passed through a thin sheet of metal faded off at the edges, and attributed this effect to scattering of the α-particles by the atoms of the metal sheet. Three years later, this scattering tendency was studied in greater detail by Geiger and E. Marsden, with the aid of scintillation counting. On directing the radiation from radium on to a thin sheet of platinum, they found that one α-particle in 8000 was scattered at an angle of 90° to the direction of incidence. Many years afterwards, the startling nature of this result was aptly described by Rutherford in the words 'It was about as credible as if you had fired a 15-inch shell at a piece of tissue paper and it came back and hit you'.

From this result, it appeared that an α-particle was unlikely to be greatly deflected in its passage through matter, but that occasionally it might receive a violent deflection that could only be accounted for by close approach to a particle of much greater mass carrying an intense positive charge. From considerations such as these, Rutherford proposed, in 1911, his theory of the nuclear atom. According to this theory, the atom consists of a small positively charged 'nucleus' in which practically all the mass of the atom is concentrated. The remaining mass is accounted for by electrons, which, in Thomson's view, are to be found in all atoms. These electrons are situated around and outside the nucleus and the number of them is such that their combined negative charge is exactly equal in magnitude to the positive charge on the nucleus.

On the basis of this theory, Rutherford proceeded to calculate the scattering behaviour of α-particles. The results of such calculations might then be compared with those obtained by experiment and the validity of the theory assessed.

7.2 Calculation of α-particle scattering

Rutherford's calculation procedure can be described with the aid of Figure 7.1. The path of an α-particle is indicated by the arrow headed curve, and is expressed in terms of Cartesian coordinates x and y. The nucleus N is regarded as sufficiently massive for any movement of it during the scattering process to be negligible. It is located at the point $x = -l$ on the x-axis. The motion of the α-particle is then that of a body repelled by a force directed from a fixed point N, the magnitude of which varies inversely as the square of the distance of the body

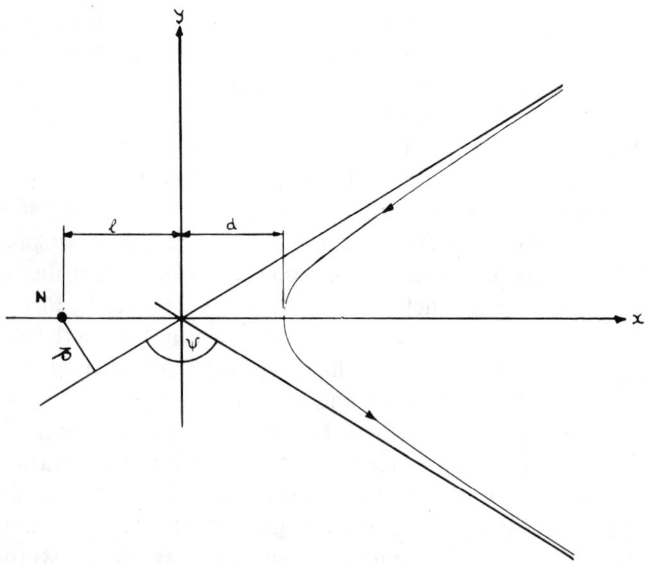

Figure 7.1 Diagram for calculation of α-particle scattering

from N, in accordance with Coulomb's law of force between charges. The problem of determining the path of such a body is a familiar one in classical mechanics, having the solution

$$\frac{x^2}{d^2} - \frac{y^2}{l^2 - d^2} = 1 \qquad (7)$$

Setting y equal to zero in this equation, the α-particle is seen to make its closest approach to the nucleus at the point $x = d$, which is indicated in Figure 7.1. Also, when x is very large, equation (7) reduces to

$$y = \frac{\sqrt{(l^2 - d^2)}}{d}\, x$$

which is the equation to two straight lines through the origin of equal and opposite slope, as shown in the Figure. If the α-particle is scattered through an angle ψ, it follows that

$$\cot(\psi/2) = \sqrt{(l^2 - d^2)}/d$$

whence

$$d = l \sin(\psi/2)$$

If the shortest distance from the nucleus to the original line of travel of the α-particle is p, then

$$p = l \cos(\psi/2)$$

Combining these last two results,

$$l + d = p\,\frac{1 + \sin(\psi/2)}{\cos(\psi/2)} \qquad (8)$$

If the velocity of the α-particle is V when remote from the nucleus and v when nearest to the nucleus, conservation of angular momentum requires that

$$pV = (l + d)v$$

or

$$v/V = p/(l + d)$$

Also, if M denotes the mass of the α-particle and E its charge, and if the charge on the nucleus is equal in magnitude to Z electronic charges e, conservation of energy leads to the relation

$$\tfrac{1}{2}MV^2 = \tfrac{1}{2}Mv^2 + \frac{ZeE}{\varkappa(l + d)}$$

where $\varkappa$ is the permittivity of free space in the c.g.s. electromagnetic or c.g.s. electrostatic systems of units. In the M.K.S. system, $\varkappa$ is 4π times the permittivity of free space. This relation may be written

$$(v/V)^2 = 1 - [b/(l+d)]$$

where

$$b = 2ZeE/\varkappa MV^2 \tag{9}$$

It may be noted that b is the shortest distance reached by an α-particle from a nucleus lying on its original line of travel. From the two above equations for v/V, it follows that

$$p^2/(l+d)^2 = 1 - [b/(l+d)]$$

whence, on eliminating $l+d$ by means of equation (8),

$$p = (b/2) \cot (\psi/2) \tag{10}$$

This treatment ignores the effect of adjacent nuclei and is therefore invalid when p is very much larger than b, so that ψ approaches zero.

It is now supposed that α-particles fall normally on to a thin sheet of material having a thickness T and containing N atoms per unit volume, so that there are NT atoms per unit area of sheet. Then the closest distance between the original line of travel of a particle and a nucleus in the material has a probability of lying between p and $p + \mathrm{d}p$ equal to

$$2\pi p \, \mathrm{d}p \,.\, NT$$

Eliminating p from this expression by means of equation (10), the corresponding probability of the particle being scattered through an angle lying between ψ and $\psi + \mathrm{d}\psi$ is

$$\tfrac{1}{4}\pi NTb^2 \cot (\psi/2) \operatorname{cosec}^2 (\psi/2) \, \mathrm{d}\psi \tag{11}$$

7.3 Experimental evidence for the nuclear atom

In order to put Rutherford's calculation of α-particle scattering to the test of experiment, Geiger and Marsden used the apparatus shown schematically in Figure 7.2. The α-rays from a radioactive source S were formed into a narrow beam by the hole H to fall normally on to a thin foil F of scattering material. The complete assembly was mounted

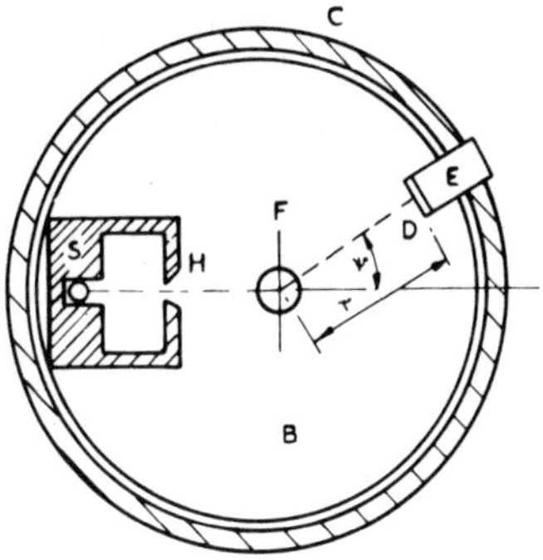

Figure 7.2 Apparatus for measurement of α-particle scattering

on a fixed base B, and was enclosed in an evacuated metal casing C. Rotation of the casing about the centre of the foil was permitted by a vacuum-tight joint between casing and base. Fixed in the casing was a form of scintillation counter, consisting of a zinc sulphide screen D attached to a microscope E, the screen being at a radius r from the centre of the foil.

If the number of α-particles counted per unit area of the screen was C when the plane of the screen made an angle ψ with that of the foil, then the number of particles scattered between angles ψ and $\psi + d\psi$ would have been

$$2\pi r \sin \psi \,.\, r\, d\psi \,.\, C$$

By expression (11), however, this number is equal to

$$\tfrac{1}{4}\pi NTb^2 \cot (\psi/2) \operatorname{cosec}^2 (\psi/2)\, d\psi \,.\, Q,$$

where Q is the total number of particles that reached the foil from the source. Equating these two expressions,

$$C = Q(NTb^2/16r^2) \operatorname{cosec}^4 (\psi/2) \qquad (12)$$

With a given foil, Geiger and Marsden varied the angle ψ from 5° to 150°, the corresponding value of $\operatorname{cosec}^4 (\psi/2)$ varying by a factor of some 250 000. The ratio of C to $\operatorname{cosec}^4 (\psi/2)$, however, was found to be very nearly constant, as required by equation (12). Maintaining the angle ψ constant, the experiments were repeated with different thicknessses of foil, and C was found to be almost exactly proportional to T, again in accordance with equation (12). As described in Section 6.4, Geiger had already measured the velocities of α-particles from a radioactive source after passing through various thicknesses of mica. By placing various thicknesses of mica between the source and the scattering foil, therefore, the α-particle velocity V could be varied in a known manner. In this way, with the same foil and at the same angle ψ, it was found that C varied nearly inversely as V^4, as equations (9) and (12) indicated. Announced in 1913, these results amply confirmed Rutherford's calculation of α-particle scattering, and thereby afforded the earliest, if not the strongest evidence for the correctness of the underlying assumption of the nuclear atom.

7.4 Nuclear charge

From equation (9), b is seen to be directly proportional to the nuclear charge Ze. With the validity of equation (12) firmly established, therefore, it now became possible to determine the nuclear charge by means of an experiment like that of Geiger and Marsden. The only additional measurement required was that of Q, obtained by direct counting of the number of α-particles incident on the scattering foil. Using foils of different elements, Geiger and Marsden obtained an approximate indication that the number Z was about half the atomic weight. In 1913, however, A. van den Broek suggested that the number Z could be more closely identified with the ordinal number of the corresponding element in the periodic table. Using a technique similar to that of Geiger and Marsden, Sir James Chadwick made, in 1920, more precise measurements of Z, obtaining for copper, silver and platinum respective values of 29.3, 46.3 and 77.4 with an error of less than 2 per cent. These elements have the respective ordinal numbers 29, 47 and 78 in the periodic table. Experimental support was thus given to van den Broek's identification of the number Z with ordinal number in this table. Further confirmation of this identity was obtained in 1922 by P. Auger and F. Perrin for argon, in 1924 by E. S. Bieler for aluminium and magnesium, and in 1925 by Rutherford and Chadwick for gold.

7.5 Displacement Law of Radioactive Change

The question remains as to how the number Z differs for the different isotopes of an element. The answer to this question is to be found in the chemistry of the radioelements. In 1905, Otto Hahn and Sir William Ramsay obtained from a thorium compound a new radioelement which, like thorium, gave rise to thorium X and thorium emanation. This new radioelement was given the name 'radiothorium'. From the same thorium compound, Hahn shortly afterwards derived yet another new radioelement which proved to be the parent of radiothorium and which he called 'mesothorium'. In 1907, being unable to perform a chemical separation of thorium and radiothorium, H. N. McCoy and W. H. Ross concluded that these two radioelements were chemically identical and that what Hahn and Ramsay had separated from thorium was not radiothorium but mesothorium. The mesothorium had then decayed to form radiothorium, and it was these two radioelements that they had finally separated. By 1910, Soddy was able to record the chemical identity of other radioelements, in particular of ionium and thorium, of mesothorium and radium, and of the radium, thorium and actinium emanations.

From radioactive decay schemes such as

$$\left.\begin{array}{l}\text{thorium} \xrightarrow{\alpha} \text{mesothorium}\\ \text{ionium} \xrightarrow{\alpha} \text{radium}\end{array}\right\}$$

and the chemical identities of the corresponding members in each scheme, Soddy was led to suggest, in 1911, that the element to which an atom of a radioelement is chemically identical becomes, on emission by the atom of an α-particle, two units lower in ordinal number in the periodic table. In order to find the effect on position in this table of β-emission, Alexander Fleck systematically investigated the chemical properties of a number of radioelements. By 1913, he had found four radioelements chemically identical with lead such that the atom of each, on emitting a β-particle, became chemically identical with bismuth. It followed that the element to which an atom of a radioelement is chemically identical becomes, on emission by the atom of a β-particle, one unit higher in ordinal number in the periodic table. The effects so established of α- and β-particle emission on correspondence with ordinal number in this table constitute the Displacement Law of Radioactive Change, put forward by Soddy in 1913.

It is now clear that many of the radioelements are not different elements in the chemical sense, but are chemically identical and differ only in atomic weight and radioactive properties. They are therefore isotopes of one another, are invariably radioactive, each with its unique half-life, and are referred to as 'radio-isotopes'. They may be denoted by the chemical symbol of one of them, such as Th for thorium or Ra for radium, with the mass number as a superscript. In this way, ionium is denoted by ^{230}Th, radiothorium by ^{228}Th, and thorium X by ^{224}Ra. The radium, thorium and actinium emanations are each radio-isotopes of the element radon and are denoted by ^{222}Rn, ^{220}Rn and ^{219}Rn respectively. The early forms of the radioactive series mentioned in Section 5.4 have now been superseded by those given in Appendix 3 for the uranium, thorium and actinium series.

7.6 Atomic number

By the Displacement Law, the change in ordinal number in the periodic table of the element to which an atom of a radio-isotope is chemically identical exactly corresponds, in both sign and magnitude, to the charge on the particle emitted. No such correspondence can be found with the mass of the particle emitted, and the inevitable conclusion follows that the chemical properties of an element are determined entirely by the number Z and that this number is therefore the same for all isotopes of an element. It further follows that, whatever the isotopic composition of an element, the number Z for the element must be an integer. From the results described in Section 7.4, it is now unquestioned that this number is identical with the ordinal number in the periodic table. In general, this numbering system arranges the elements in order of increasing atomic weight. Exceptions are elements for which Z is equal to 19, 28, 53 and 91, the corresponding atomic weight in each case being slightly less than that of the element of next lower value of Z. The number Z is now known as the 'atomic number', and is listed for various elements in the third column of the table in Appendix 2.

It follows that the nucleus of the hydrogen atom carries a positive charge equal in magnitude to the electronic charge, or one unit positive charge. The atom is completed by the addition of one electron outside the nucleus. Similarly, the helium atom consists of a nucleus carrying two unit positive charges and two electrons outside the nucleus. The α-particle, being a helium atom that has lost two electrons, is thus seen to be simply a helium nucleus. Since the number of electrons outside the nucleus is about half the atomic weight, it follows that the contribution of these electrons to the mass of the atom is very small. The determination of the nuclear charge from Rutherford's calculation of α-particle scattering thus leads once more to a result consistent with the assumption of the nuclear atom on which the calculation is based.

In the Rutherford nuclear atom the electrons were envisaged as disposed about the nucleus in the form of a mist. No mechanism was proposed for overcoming the mutual electrostatic attraction between the electron and the nucleus. This deficiency was overcome when, in 1913, Rutherford's model of the atom was extended by N. H. D. Bohr. In this extension the electron, of charge e and mass m, revolved at speed v in a circular orbit of radius r centred on the nucleus of atomic number Z. Equating the electric and centrifugal forces on the electron,

$$Ze^2/\varkappa r^2 = mv^2/r$$

where $\varkappa$ has the same meaning as in Section 7.2. This equation relates r and v without fixing either of these two quantities. In a further development of the theory, Bohr obtained an expression for r which led to the conclusion that the electrons could only revolve about the nucleus in 'permitted' orbits.

7.7 Dimensions of the atom

Since the α-particle is itself an atomic nucleus, the distance b given by equation (9) should be of the same order as the diameter of a nucleus. Adopting the M. K. S. system of units and taking

$$\begin{aligned} Z &= 25 \\ V &= 1.5 \times 10^7 \text{ m/sec} \end{aligned}$$

and since

$$\begin{aligned} e &= 1.602 \times 10^{-19} \text{ coulombs} \\ E/M &= 4.823 \times 10^7 \text{ coulombs/kg} \\ K &= 1.113 \times 10^{-10} \text{ coulombs}^2\ \text{sec}^2/\text{kg.m}^3 \end{aligned}$$

equation (9) gives

$$\begin{aligned} b &= 1.54 \times 10^{-14} \text{ m} \\ &= 1.54 \times 10^{-12} \text{ cm} \end{aligned}$$

It is instructive to compare this result with the diameter of an atom complete with extranuclear electrons. This diameter is given approximately by the cube root of the volume per atom of an element in which the atoms are closely packed, such as a metal. From Section 2.5,

$$\text{volume/atom} = \mathscr{A}/\varrho N_0$$

where ϱ is the density of the element. In the case of metals, $\mathscr{A}/\varrho$ has a minimum value of about 6.5 cm^3/g when the atomic number ranges from 25 to 28, close to that used to obtain the above result for b. The diameter of the atom is then approximately

$$\sqrt[3]{\left(\frac{6.5}{6.022 \times 10^{23}}\right)} = 2.2 \times 10^{-8} \text{ cm}.$$

On comparing this result with that obtained for b, the diameter for the complete atom is seen to be some 10 000 times as great as the diameter of the nucleus. Once again, therefore, Rutherford's calculation of α-particle scattering yields information in accordance with its underlying assumption of the nuclear atom.

8

Artificial transmutation of the nucleus

8.1 Scattering of α-particles by light atoms

The idea of the nuclear atom arose from the scattering of α-particles by comparatively heavy atoms. As equation (9) shows, however, the α-particles will approach more closely to the nuclei of lower atomic number in the lighter atoms. Any deviation in their scattering behaviour with closeness of approach was therefore to be looked for in the passage of α-particles through elements of low atomic weight. The deviation expected was one from the Coulomb law of force between the α-particle and a nearby nucleus, a law implicit for heavy nuclei from the same experiments that confirmed the idea of the nucleus itself. In the event, a deviation was found that far exceeded all expectations. It amounted in fact to the first artificial transmutation of an element, a realisation of the alchemists' dream which scientific thought had finally deemed impossible.

It had now become clear that an α-particle was gradually slowed down in its passage through matter and finally stopped. The accompanying loss in its energy was believed to be largely due to its ionisation of the atoms in its path and hardly at all to scattering by them. Convincing evidence for this belief was immediately forthcoming with C. T. R. Wilson's demonstration in 1911 that the path of a single ionising particle could be rendered visible in a cloud chamber. An ionising particle entering the chamber during the brief period of an expansion was revealed by the trail of ions in its path acting as condensation nuclei. A line of fine droplets, or 'cloud track', was thus formed, exactly tracing the path of the ionising particle as a white line on a dark background. With the aid of two cameras having their axes at right angles to one another and at 45° to the axis of the cloud chamber, two photographs of the cloud track could be obtained from which it was possible to deduce the path of the ionising particle in three dimensions.

The cloud tracks produced by α-particles from a radioactive source just within the cloud chamber were found to consist of straight lines radiating from the source. These tracks were nearly all of the same length, corresponding to the maximum range of the α-particles. On approaching their furthest distance from the source, however, the tracks were liable to be deflected, as by collision with the nuclei of oxygen or nitrogen in the air. A typical cloud track revealing a large deflection of an α-particle through collision with an atomic nucleus is shown in Figure 8.1. Just before the collision, the α-particle is travelling in the same general direction as the other particles, with velocity V. Immediately after the collision, the α-particle travels with velocity v in a direction inclined at an angle ψ to its original line of travel. The nucleus recoils in a direction inclined at an angle φ to this line with velocity u. If M, m denote the respective masses of the α-particle and nucleus, and if momentum and energy are conserved in the collision, then

$$\left.\begin{aligned} (v/V)\cos\psi + (m/M)(u/V)\cos\varphi &= 1 \\ (v/V)\sin\psi - (m/M)(u/V)\sin\varphi &= 0 \\ (v/V)^2 + (m/M)(u/V)^2 &= 1 \end{aligned}\right\} \qquad (13)$$

It follows that

$$\left.\begin{aligned} m/M &= \sin\psi/\sin(2\varphi + \psi) \\ v/V &= \sin\psi/\sin(\varphi + \psi) \\ u/v &= \sin(2\varphi + \psi)/\sin\varphi \end{aligned}\right\} \qquad (14)$$

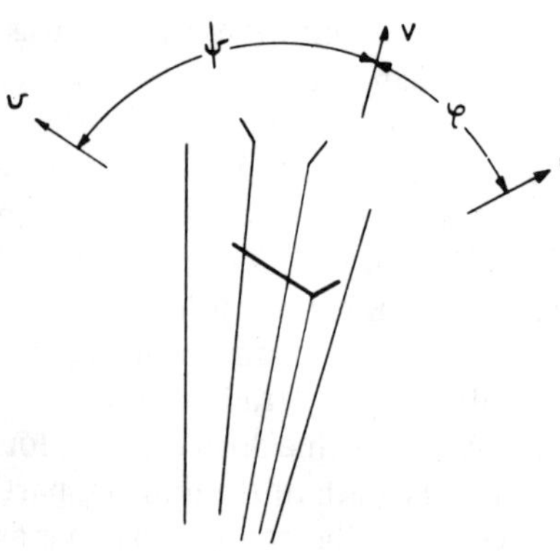

Figure 8.1 Typical cloud track of collision between α-particle and atomic nucleus

By 1923, P. M. S. Blackett had measured the angles φ and ψ for several branched cloud tracks in various gases. These angles were found to be in accordance with the first of equations (14). The conservation of energy and momentum in collisions between α-particles and atomic nuclei was thereby confirmed.

By this time, also, Blackett had been able to deduce range–velocity relationships for the recoil nuclei of nitrogen, hydrogen, helium and argon – for the velocity v of the α-particle just after collision was given by its range as determined by the length of its track beyond the branching point and the gas in question. The third of equations (14) then gave the velocity u of the recoil nucleus just after collision, the length of its track being converted to the corresponding range as for α-particles. The chief difficulty in these determinations arose from the very short ranges of the α-particles at the point of collision. At these ranges, which rarely exceed 10 mm, not only Geiger's Law but Geiger's experimental technique itself breaks down, due to the scatter in velocity of the α-particles issuing from the larger thicknesses of mica, resulting in a broad diffuse beam in a magnetic field.

To overcome this difficulty, Blackett made use of a generalised form of equation (10) that took into account the finite mass of the recoil nucleus. The precise nature of the generalised equation is not of immediate interest but dimensional analysis shows that it will be of the type

$$p = bF(\psi, m/M, Ze/E)$$

where the symbols p, b, Z, e and E have the same meaning as in equations (9) and (10). The probability of a collision occurring in a distance $\mathrm{d}T$ such that the α-particle is deflected through angles between ψ and $\psi + \mathrm{d}\psi$ is then, analogous with expression (11),

$$\pi N \mathrm{d}T\, b^2 \frac{\mathrm{d}F^2}{\mathrm{d}\psi}\,\mathrm{d}\psi$$

$$= \frac{K}{V^4}\,\mathrm{d}T \frac{dF^2}{\mathrm{d}\psi}\,\mathrm{d}\psi \qquad (15)$$

where

$$K = 4\pi N(ZeE/\varkappa M)^2$$

Blackett's analysis now proceeds somewhat as follows. If the α-particle is travelling through air, then the residual range just before collision can be written

$$T = kV^n$$

where the index n is to be determined and k is a constant. Likewise, the range immediately after collision is

$$R = kv^n$$

Also, the first and second of equations (14) give

$$v/V = f(\psi, m/M)$$

so that

$$T = Rf^{-n}$$

With these relationships expression (15) becomes

$$K(k/R)^{4/n} f^{4-n}\,\mathrm{d}R \frac{\mathrm{d}}{\mathrm{d}\psi} F^2\,\mathrm{d}\psi$$

so that the probability of a collision for which

$$\left.\begin{array}{l}\psi_1 < \psi < \psi_2 \\ R_1 < R < R_2\end{array}\right\}$$

is

$$k^{4/n}K \int_{\psi_1}^{\psi_2} f^{4-n} \frac{\mathrm{d}}{\mathrm{d}\psi} F^2\,\mathrm{d}\psi \int_{R_1}^{R_2} R^{-4/n}\,\mathrm{d}R$$

From a large number of collisions for which ψ lies between prescribed values ψ_2 and ψ_1, the number of collisions for which R lies between $j\Delta R$ and $(j + 1)\Delta R$ is proportional to

$$\int_{jR}^{(j+1)\Delta R} R^{-4/n}\,\mathrm{d}R = (R^{-4/n})_{\mathrm{av}}\Delta R$$

where j has one of the values 0, 1, 2, . . . and the suffix av denotes an average value over the $(j + 1)^{\mathrm{th}}$ interval ΔR. If the collisions recorded experimentally for which ψ lies between the prescribed values are subdivided into groups such that each group corresponds to one of the intervals ΔR, then the number in each such group is proportional to $(R^{-4/n})_{\mathrm{av}}$ and may be plotted against the corresponding average value of R. From the resulting plot, the index n can be determined. In this way, from measurements made on 1524 branched cloud tracks in air, Blackett concluded that the index n was 3/2.

8.2 The first artificial transmutation

The knowledge gained on the general behaviour of α-particles in gases was now applied to a study of their scattering by the lightest nuclei, for the extreme closeness of approach of particle to nucleus in scattering of this kind would reveal any departure there might be from the Coulomb law of force between them. One obvious choice of scattering element was hydrogen, and in 1914, on causing α-particles to pass through hydrogen, E. Marsden recorded scintillations on a zinc sulphide screen placed well beyond the reach of the α-particles. It was concluded, rightly as it turned out, that these scintillations were produced by hydrogen nuclei knocked on by the α-particles.

In 1919, Rutherford reported the results obtained from a more elaborate version of Marsden's experiment, the apparatus used being as shown diagrammatically in Figure 8.2. An α-particle source S was mounted in a rectangular vessel V which could be evacuated or filled with a suitable gas. A small hole at one end of the vessel was covered by a window W of thin silver foil of stopping power equal to the maximum range of the α-particles. Immediately beyond this window was a zinc sulphide screen D, the gap between the two being sufficient for the insertion of sheets of absorbing material. A strong magnetic field bent away the β-rays emitted by the source so that they did not reach the screen.

With the vessel evacuated, only a few scintillations were observed. They only ceased when the stopping power of the window and absorbing sheets reached 280 mm or so, a value well in excess of the maximum range of the α-particles. When the vessel was filled with hydrogen, the number of scintillations increased enormously, but again a stopping power of some 280 mm was required in order to prevent the particles that produced the scintillations from reaching the screen. It was concluded, as it had been previously by Marsden, that these particles were hydrogen nuclei, the scintillations obtained with the vessel evacuated being attributed to hydrogen given off by the materials of the apparatus. Support for this conclusion was obtained from the deflection of the particles in a magnetic field, which gave the charge-to-mass ratio of the particles in terms of their velocity. An

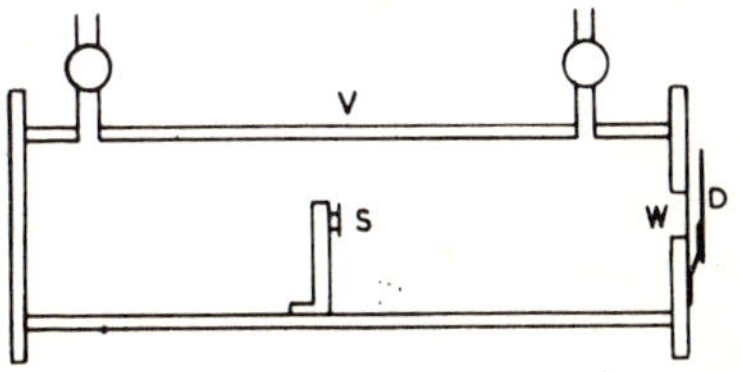

Figure 8.2 Rutherford's apparatus for projection of hydrogen nuclei by α-particles

estimate of this velocity was obtained from equations (13), which, in the case of a head-on collision, reduce to

$$u = 2\,V/[(m/M) + 1] \tag{16}$$

Since, by Geiger's Law, the velocity of an α-particle falls only slowly with quite large reductions in range, the velocity V would not generally differ greatly from the initial velocity of the α-particle, as deduced from Geiger's Law and the known maximum range of α-particles from the particular source. On assuming a value for the ratio m/M, therefore, the initial velocity u of the knocked-on particle could readily be found. With m/M set equal to the ratio of the mass of the hydrogen atom to that of the α-particle, a velocity u was obtained that, applied to the results of the magnetic deflection experiment, gave a charge-to-mass ratio about equal to that for the hydrogen nucleus. That the knocked-on particles were in fact hydrogen nuclei was thus clearly demonstrated.

On replacing the hydrogen in the vessel by oxygen or carbon dioxide no scintillations were observed on the zinc sulphide screen. On filling the vessel with nitrogen, however, a large number of scintillations appeared, similar in character to those obtained with hydrogen. As in the case of hydrogen, a stopping power very much greater than the maximum range of the α-particles was required in order to prevent these scintillations from appearing. Rutherford was thus led to the view that these scintillations were produced by hydrogen nuclei that had been knocked out of the nitrogen nuclei by the α-particles. By 1922, Rutherford and Sir James Chadwick had established that the stopping power that was just sufficient to prevent the appearance of scintillations in the case of nitrogen was about 400 mm as opposed to the corresponding figure of 280 mm in the case of hydrogen. Any possibility that the scintillations in the nitrogen case might be due to a hydrogen impurity was thus disposed of.

In a later form of the apparatus depicted in Figure 8.2, the source S was surrounded by a close fitting cover, having an aperture in the end nearest the screen D. Across this aperture, thin foils of various materials could be placed which, with the vessel V evacuated, could be examined for the production of scintillations on the screen. In this way, Rutherford and Chadwick had, by 1924, examined all elements of atomic number from 5 to 19. Of these, all except carbon and oxygen were found to produce scintillations after insertion of absorbing sheets of stopping power greater than 320 mm – sufficient to exclude scintillations due to the presence of hydrogen.

At the same time, in the case of a few of these elements, measurements were made of the deflection in a magnetic field of the particles producing the scintillations. For this purpose a brass plate was interposed between the source and the screen, being at right angles to the screen and to one side of the normal to the screen through the source. The particles emitted from the element under examination then formed a shadow on the screen of the edge of the plate. By applying a magnetic field parallel to the screen and plate, first in one direction and then in the other, the border of the shadow was displaced in a direction at right angles to the plane of the plate, so that the deflection of the particles could be computed. This deflection at once gave the charge-to-mass ratio of the particles in terms of their velocity. In the case of hydrogen the velocity of the particles and the stopping power that just prevented the appearance of the scintillations had already been determined and in the case of any other element the velocity could be deduced by proportioning the cube of it to the corresponding stopping power on the assumption that the particles were indeed hydrogen nuclei and that a law of the form of Geiger's applied. In this way, the charge-to-mass ratio for the particles was found to be of the same order as that for the nucleus of hydrogen and the identity of the particles with hydrogen nuclei was thereby confirmed. A rough estimate could now be made of the kinetic energy of the particles and this, in the case of aluminium, was found to be about 40 per cent greater than the initial energy of the α-particles used to produce it. It was now clear that Rutherford's original experiment with nitrogen had given nothing less than the first demonstration of transmutation of the atomic nucleus by an external agent. Transmutation of this kind is said to be 'artificial'.

The emission of hydrogen nuclei by many different elements on bombardment by α-particles pointed to the hydrogen nucleus as a universal constituent of all atomic nuclei. By this time, therefore, the hydrogen nucleus had acquired a special significance which was marked by giving it the name 'proton'.

The precise manner in which an α-particle interacted with an atomic nucleus to eject a proton was at first in doubt. Either the affected nucleus merely lost a proton, so that the α-particle continued on its way after the interaction, or the nucleus not only lost a proton but acquired an α-particle as well. This uncertainty was resolved in 1925 when Blackett obtained cloud tracks of the ejection of protons by nitrogen nuclei when struck by α-particles. These were of the branched form shown in Figure 8.1. The track of the interacting nucleus closely resembled that of the recoil nucleus in Figure 8.1, whilst the track of the ejected proton was rather longer and very much finer than that of the α-particle after collision as shown in the figure. The absence of a fourth track radiating from the branching point showed that the α-particle did not continue on its way after the interaction but was absorbed by the interacting nucleus.

8.3 Fission of lithium by accelerated protons

That the success of Rutherford and Chadwick in achieving nuclear disintegration should have been limited to elements of low atomic number is to be expected for, as equation (9) shows, it is in just such elements that the bombarding particle would appear to approach the target nucleus most closely. The closeness of this approach, however, is also seen to depend on the bombarding particle itself and, with the capabilities of the α-particle apparently exhausted, the need was now felt for a more effective one. The twin requirements of high kinetic energy and low nuclear charge focused attention on the artificially accelerated proton and, in 1932, bombarding particles of this type were successfully used by Sir John D. Cockcroft and E. T. S. Walton in the fission of lithium.

The apparatus used by Cockcroft and Walton is outlined in Figure 8.3. Protons from a hydrogen discharge tube entered the apparatus by the tube H. They were accelerated in the gaps between electrodes T by means of an electric potential capable of reaching 600 000 volts which was achieved by charging capacitors in parallel and discharging them in series. The electrodes were of thin-walled steel tube and fitted at their ends with steel rings R. They were supported on steel plates P separated by glass cylinders C, each cylinder being 356 mm in diameter and 914 mm in length. The apparatus was evacuated by pumping from the outlet G. After passing through a thin mica leaf window W, the resulting proton beam struck a layer of lithium evaporated onto a thin mica film F, set at an angle of 45° to the beam. Two zinc sulphide screens D were set parallel to the proton beam one on either side of the film and at an angle of 45° to it. Any particle reaching the screens from the film did so through mica sheets S. These sheets had a stopping power of 14 mm, sufficient to prevent the highest energy protons used from reaching the screens. The screens were viewed through microscopes E.

When the accelerating potential reached about 125 000 volts, scintillations appeared on the two screens simultaneously and

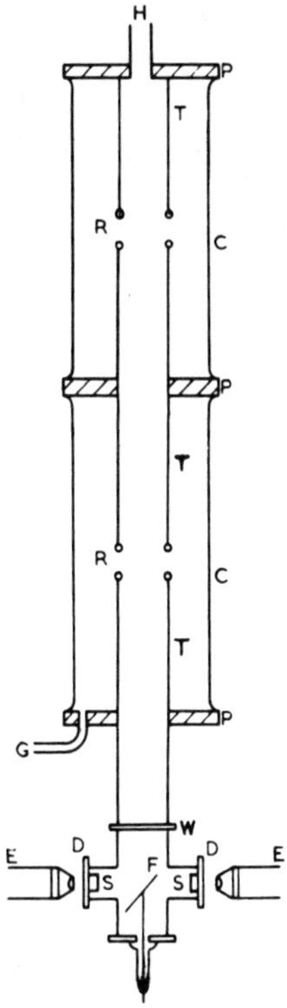

Figure 8.3 Cockroft and Walton's apparatus for fission of lithium by protons

increased in number as the voltage was further raised. These scintillations were identical to those produced by the highest energy α-rays from natural radioelements and evidence was forthcoming that they occurred coincidently on both screens. It thus appeared that the ^{7}Li isotope of atomic number 3 occasionally captured a proton, and that the resulting nucleus of atomic number 4 and mass number 8 then broke up into two α-particles. Support for this conclusion was provided by cloud chamber photographs taken by P. I. Dee and E. T. S. Walton a year later. Two tracks of the density associated with α-particles were seen to have emerged in opposite directions from a lithium target bombarded by protons. These tracks were of equal length after correcting for the stopping power of the lithium and underlying mica. This result is consistent with the conservation of momentum of a stationary nucleus splitting into two equal particles.

8.4 Nuclear reactions

Rutherford's transmutation of nitrogen and Cockcroft and Walton's fission of lithium are both particular examples of what are generally termed 'nuclear reactions'. In such reactions, it seems reasonable to suppose that the sum of the atomic numbers will be the same for the particles giving rise to the reaction as for those resulting from it. Furthermore, Aston's whole number rule suggests that the same will be true of the mass numbers. In fact, denoting atomic numbers by subscripts, nuclear reactions can be represented by equations of the form used to describe chemical reactions and written, in the case of the two examples already quoted,

$$^{14}_{7}\text{N} + {}^{4}_{2}\text{He} \rightarrow {}^{1}_{1}\text{H} + {}^{17}_{8}\text{O}$$
$$^{7}_{3}\text{Li} + {}^{1}_{1}\text{H} \rightarrow {}^{4}_{2}\text{He} + {}^{4}_{2}\text{He}$$

Just as the number of atoms of each element balance on the two sides of the equation for a chemical reaction, so do the sum of the atomic numbers and the sum of the mass numbers balance in the equation for a nuclear reaction. Since the atomic number subscript conveys no information additional to that already given by the chemical symbol, it may be omitted and use made instead of a simpler scheme devised by W. Bothe in 1935. According to this scheme, the two nuclear reactions quoted above would be formulated as

$$^{14}\text{N}(\alpha,\text{p})^{17}\text{O}$$
$$^{7}\text{Li}(\text{p},\alpha)^{4}\text{He}$$

where the first symbol denotes the target nucleus, the last symbol the residual nucleus and the first and second bracketed symbols the incident and ejected particle respectively.

As the cloud tracks of the fission of ^{7}Li by protons showed, momentum is conserved in nuclear reactions, but energy, as in chemical reactions, may either be absorbed or liberated. In the fission of a ^{7}Li nucleus by a proton, for instance, energy is liberated to the extent of that carried away by the ejected α-particles less that brought in by the incident proton. Cockcroft and Walton measured the range of the ejected α-particles with the apparatus shown in Figure 8.3. A mica window was placed in the position previously held by the zinc sulphide screen D and the screen was repositioned immediately outside that window. By inserting layers of mica between screen and window, the α-particle range was found to be 84 mm, corresponding to a particle energy of 8.65 MeV. The proton carries a charge equal in magnitude to a single electronic charge so that the accelerating potential of 125 000 volts, the lowest at which Cockcroft and Walton obtained evidence of fission, corresponds to an incident proton energy of only 0.125 MeV. The energy liberated by the reaction therefore amounted to 17.2 MeV.

By this time, however, an alternative method of calculating the energy of a nuclear reaction was available, which demanded only a knowledge of the masses of the various atomic nuclei. But most of these masses had already been determined with considerable accuracy by the mass spectrograph and only the validity of the method remained to be demonstrated. The method itself rested on a proposition derived by Albert Einstein in 1905 from his theory of relativity. According to Einstein, a system undergoing a change dE in its energy E, of whatever form, suffers a proportional and similarly directed change dm in its mass m such that

$$\mathrm{d}m = (1/c^2)\,\mathrm{d}E$$

where c denotes the velocity of light. It follows that

$$\mathrm{d}E/\mathrm{d}m = 931.48 \text{ MeV/amu} \qquad (17)$$

A consequence of this relationship is that the mass of a particle depends on its velocity, so that the mass m of a particle and its velocity v are connected by the equation

$$m = m_0/\sqrt{1-(v/c)^2}$$

where m_0 is the mass of the particle when at rest, known as the 'rest mass'. On acquiring the velocity v, therefore, the corresponding gain in energy, i.e. in kinetic energy, is

$$\begin{aligned}(m - m_0)c^2 &= m_0c^2\left[\frac{1}{\sqrt{[1-(v/c)^2]}} - 1\right]\\ &= m_0c^2\,[1 + \tfrac{1}{2}(v/c)^2 + \tfrac{3}{8}(v/c)^4 + \ldots]\\ &\approx \tfrac{1}{2}m_0v^2\end{aligned}$$

when v is small compared with c, so that under this condition the result tends to that given by classical mechanics.

In the case of a nuclear reaction the masses of interest are the 'nuclear weights', which are the isotopic weights less the masses in amu of the electrons associated with the non-ionised atoms. However, since the reaction leaves the sum of the atomic numbers and therewith the sum of the included electrons unchanged, the reactants are equally well represented by their isotopic as by their nuclear weights. For the sake of consistency, isotopic weights are invariably corrected to their rest masses.

In applying the relationship (17) to a nuclear reaction, the system is regarded as consisting initially of the incident particle and target nucleus at rest. The kinetic energy of the incident particle at the instant of reaction is then the energy absorbed, or gained, by the system. In its final state, the system is considered to be formed by the ejected particle and residual nucleus at rest, their combined kinetic energy at the instant of reaction being energy liberated, or lost, by the system. The relation (17) may now be applied on writing the nett gain in energy by the system for dE and the nett gain in isotopic weight of the system for dm.

In 1930, Rutherford, Chadwick and C. D. Ellis showed that, in the natural decay of uranium to lead, their isotopic weights, as well as those of the particles radiated, were likely to conform with the measured particle energies as required by the relation (17). Cockcroft and Walton found fair agreement between this relation and the results of their experiment on the fission of ^{7}Li by protons.

In 1933 K. T. Bainbridge checked this reaction against the relationship (17) using more accurate isotopic weights. In terms of the isotopic weights listed in Appendix 2, the calculation proceeded as shown in Table 8.1.

Table 8.1

Reacting particle	Formula	Isotopic Weight (amu)
Target nucleus	^{7}Li	7.0160
Incident proton	^{1}H	1.0078
Ejected α-particle	^{4}He	4.0026
Residual nucleus	^{4}He	4.0026

From the table

$$\begin{aligned}\text{nett gain in energy} &= 931\,(4.0026 + 4.0026 - 7.0160 - 1.0078)\\ &= -17.3 \text{ MeV}\end{aligned}$$

This calculated value agrees very closely with the measured one already noted. From agreement such as this, Bainbridge obtained strong support for the application of the Einstein mass–energy relationship to nuclear reactions. Subsequent measurements of energy and mass in other nuclear reactions have amply confirmed this relationship so that its validity, as also the reality of nuclear reactions themselves, is now unquestioned.

A reaction in which energy is liberated may be triggered off by incident particles of much lower energy, as is the case with the fission of ^{7}Li by protons. On the other hand, a reaction in which energy is absorbed cannot take place until the incident particles have attained a greater energy, known as the 'threshold energy'. The excess of the threshold energy over that absorbed in the reaction is equal to the sum of the energies of the ejected particle and residual nucleus.

8.5 The cyclotron

Progress in the artificial production of nuclear reactions rests very largely on the means used to generate and project the incident particles. As already noted, the α-particle provided the earliest example of an incident particle, being first used as such by Rutherford. For the generation and projection of this particle, however, Rutherford relied on the natural means of radioactivity. The energy of the particle was thereby limited but the way to higher energies was pointed by Cockcroft and Walton, who not only achieved yet another artificial nuclear transmutation but who first did so with incident particles that were artificially both generated and projected. The method used by Cockcroft and Walton was capable of yielding incident particles of energy up to 1 or 2 MeV. Since that time, the continual urge towards incident particles of ever higher energy has prompted several developments, but of these the most striking is probably the cyclotron.

Invented by E. O. Lawrence in 1932, the cyclotron has since been developed to yield incident particles of energy approaching 400 MeV. A diagrammatic representation of a cyclotron is provided by Figure 8.4. A flat circular metal box is cut into semicircular halves D separated by a gap. Because of their shape, the two halves are referred to as 'dees'. They are completely surrounded by a closed casing C, containing gas at low pressure. This casing is placed between the poles of a magnet M, coaxial with the pair of dees. An alternating potential from a source A is applied between the dees, which then act as electrodes.

The generation of the incident particles, as distinct from their projection, rests on a discovery made by Thomas Edison in 1883. On sealing a metal filament into a highly evacuated glass bulb, Edison heated it by connecting its two ends outside the bulb to the terminals

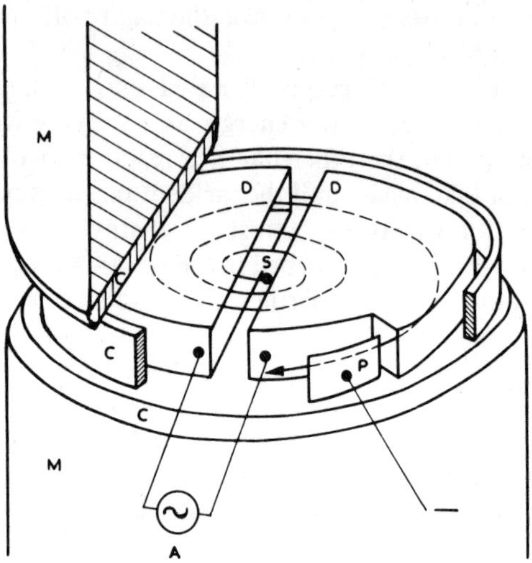

Figure 8.4 The cyclotron

of a battery. Also sealed into the bulb was a metal electrode which could be connected outside the bulb through a milliammeter, to either terminal of the battery. Although the battery voltage was well below that required for the production of cathode rays, current flowed through the electrode when connected to the positive terminal, but not when connected to the negative one. Even with the electrode connected to the positive terminal, however, no current passed through it when the filament was cold. It thus became clear that negative electricity was given off by a heated filament, a phenomenon referred to as 'thermionic emission'. In 1899, supposing this emission to be in the form of charged particles, Sir J. J. Thomson determined their charge-to-mass ratio by means of electric and magnetic fields, and thus identified the emission as one of electrons.

In the cyclotron, thermionic emission takes place from an electrically heated filament S placed at the centre of the gap between the dees. The resulting stream of electrons causes ionisation by collision of the gas contained in the casing, thereby generating the required incident particles. If the gas is hydrogen, for instance, the generated particles are protons or if it is helium, they are α-particles.

A particle thus generated will move from the filament S towards the dee which happens at that instant to have a negative potential, and in so doing will acquire a certain velocity. The particle will maintain this velocity within the dee, but due to the field of the magnet M it will now move in a circular path and will thus arrive once more at the open end of the dee. If by this time the potential on the dees has been reversed, then the particle will be accelerated across the gap to enter the other dee at an increased velocity. Its subsequent path within this dee will again be circular but its higher velocity will cause the radius of the path to be greater than before. In this way, the particle follows within the dees a continuous spiral path originating from the source S, its velocity increasing each time it crosses the gap between the dees. Eventually, the particle reaches the periphery of the dees, travelling at a very great velocity. It can now be brought out of the dees by means of a deflecting plate P charged to a high negative potential and thus directed through a thin mica window on to any desired target material.

If, over any length of its path of radius of curvature r, the particle, supposedly of mass M and carrying charge E, is travelling with a velocity v, then, on equating magnetic and centrifugal forces on the particle, the relation

$$BEv = Mv^2/r$$

is obtained, where B is the flux density of the field due to the magnet M. The time taken by the particle in tracing the complete semicircle of radius r within one of the dees is then

$$\pi r/v = \pi M/BE$$

and is clearly the same for every semicircle so traced. It follows that a constant value can be found for the frequency of oscillation of the potential on the dees so that the changes in potential keep in phase with the movements of the particle from one dee to the other.

9

The structure of the nucleus

9.1 Discovery of the neutron

In its use for nuclear bombardment, the α-particle, naturally generated and projected, has been seen to have faced a decline due to the urge for more effective incident particles. That such particles must possess increased energy, reduced electric charge, or both, has already been recognised, as exemplified by Cockroft and Walton's use of the accelerated proton. As far back as 1920, however, Rutherford had conceived the idea of obtaining a particle of even zero charge by causing an electron to combine with a proton. At about this time, W. D. Harkins also envisaged such a particle and gave to it, in 1921, the name 'neutron'. Thereafter numerous attempts at the formation of the neutron met with failure. Yet it was that most primitive of nuclear projectiles, the naturally occurring α-particle, that gave rise to the neutron as at first detected. Once discovered, the neutron was to prove not only a far more effective projectile than the α-particle that brought it to light but also the means by which the nucleus was finally forced to unleash its store of energy on an unprecedented scale.

The first step towards discovery of the neutron, though unrecognised as such at the time, was taken in 1930. In that year, W. Bothe and H. Becker observed that lithium, boron and notably beryllium, when exposed to the α-rays from polonium, emitted a radiation detectable by a Geiger counter. By causing the radiation to pass through an absorber, they showed it to be of exceptional penetrating power, and classified it as a γ-type. On studying this radiation in 1932, F. Joliot and Irene Curie-Joliot found that it was readily absorbed by hydrogenous materials such as paraffin wax and water, at the same time causing high energy protons to be emitted. In the same year, the radiation was passed through a cloud chamber by N. Feather, and, although no track was seen of the radiation itself, its presence was revealed by the tracks of nitrogen nuclei recoiling from its path. Sir James Chadwick was now led to the view that the radiation consisted of particles, that these particles were uncharged to account for their high penetrating power and, furthermore, that the particles were neutrons (^1_0n) emitted as a result of a reaction such as

$$^9_4Be + ^4_2He \rightarrow ^{12}_6C + ^1_0n$$

In order to confirm his ideas, Chadwick carried out a series of experiments with the aid of the apparatus shown diagrammatically in Figure 9.1. The source vessel S was evacuated at the outlet G and contained an α-particle emitter P in the form of polonium on a silver disc. The α-particle impinged on a beryllium disc B, causing radiation to be emitted in the direction of an ionisation chamber I. This chamber was fitted with a window W of aluminium foil, the electrode E being connected through an amplifier to an oscillograph.

When the source vessel was placed in front of the ionisation chamber, the number of counts at once increased from 7 per hour to 4 per minute, being then unaffected by the interposition of a 20 mm thickness of lead. The radiation reaching the chamber was thereby shown to be highly penetrating, as Bothe and Becker had already observed. When the lead was replaced by a 2 mm thickness of paraffin wax, the number of counts increased greatly. By placing screens of aluminium between the wax and the ionisation chamber, the particles causing this increase were found to require a stopping power of 400 mm to prevent their reaching the chamber. The oscillograph deflections were of the same size as those due to protons of similar range. The radiation from the paraffin was thus shown to consist of protons, as the Joliots had concluded earlier.

On the assumption only that the radiation emitted by the beryllium did consist of uncharged particles of some kind, Chadwick proceeded

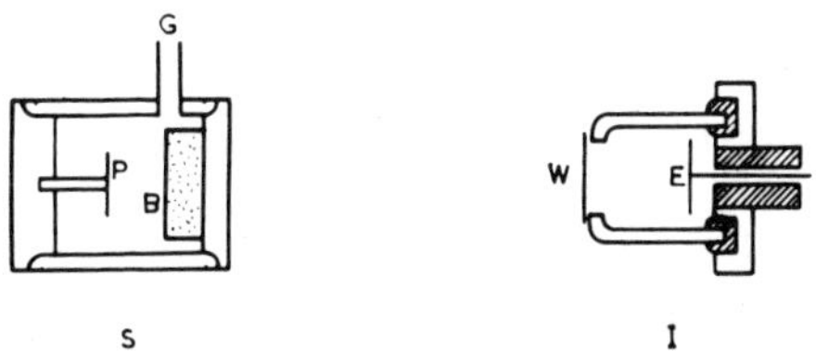

Figure 9.1 Chadwick's neutron source and ionisation chamber

to estimate their mass. Protons of mass M_p, ejected from paraffin wax by the uncharged particles of mass M_n and velocity v_n, were assumed to acquire their maximum velocity v_p as the result of head-on collisions, so that, from equation (16),

$$v_p = 2v_n/[(M_p/M_n) + 1]$$

Similarly, nitrogen nuclei of mass M_N, recoiling from collisions with the uncharged particles in a cloud chamber, were taken to have a maximum velocity

$$v_N = 2v_n/[(M_N/ M_n) + 1]$$

Eliminating the velocity v_n from these two equations,

$$v_p/v_N = (M_N + M_n)/(M_p + M_n)$$

From the stopping power needed to cut off the protons, the velocity v_p was estimated to be 11.0 per cent of the velocity of light and from the maximum length of the cloud tracks of the recoil nitrogen nuclei, the corresponding value of v_N was found to be 1.56 per cent. Taking the masses M_p and M_N as 1.0 and 14.0 amu respectively, the mass M_n became 1.15 amu close to the mass of a proton.

A further calculation of this kind was now necessary for collisions with a different pair of nuclei, so that a second value could be obtained for the mass of the uncharged particle. Identity of this value with the previous one would then verify the assumption that the radiation emitted by the beryllium did indeed consist of uncharged particles. Accordingly, the number of ion pairs created by a nucleus of nitrogen and by one of argon on recoiling from collision with one of the uncharged particles were determined experimentally and found to have maximum values agreeing well with the energies that these nuclei would acquire as a result of head-on collision with a particle of mass M_n and velocity v_n as already estimated. These determinations, although approximate, were nevertheless sufficient to justify Chadwick's view that the radiation leaving the beryllium not only consisted of uncharged particles but that these particles were, in fact, neutrons.

Having obtained a first approximation to the mass of the neutron, Chadwick now used it to derive a second approximation from the Einstein mass–energy relationship. It was already known that boron, when bombarded by α-particles, emitted a radiation similar in every respect to that emitted under identical circumstances by beryllium. The radiation emitted by boron in this way was thus assumed to consist of neutrons. Under these conditions, however, Rutherford and

Chadwick had established emission of a proton and application of the mass–energy relationship suggested that it was the ^{10}B isotope that took part in that emission. The neutron radiation was therefore attributed to the reaction $^{11}B(\alpha,n)^{14}N$. Referring to Figure 9.1, the beryllium disc B was replaced by a graphite plate covered with powdered boron. Protons were ejected from paraffin wax in front of the source chamber S and the stopping power that suppressed them was measured as before. This stopping power gave the maximum velocity of the protons and, on the assumption of a head-on collision, the corresponding energy of the impinging neutrons. The energy of the α-particles emitted by polonium being known, the energy of the recoiling ^{14}N nucleus was determined so as to ensure conservation of momentum on the assumption that α-particle, ^{14}N nucleus and neutron moved along the same straight line. With the nett gain in energy of the reaction thus determined, the nuclear weights of ^{11}B, ^{4}He and ^{14}N then gave a value for the mass of the neutron between 1.005 and 1.008 amu. Various nuclear reactions involving neutrons have since been analysed in this way, and the mass of the neutron thereby more accurately determined as

1.008665 amu

9.2 Constituent particles of the nucleus

With the discovery of the neutron, ideas on the structure of the nucleus underwent an immediate change into a form that is now beyond question. Although the atomic nucleus was not established until 1913, ideas on its structure can be traced to a hypothesis put forward by W. Prout nearly a hundred years earlier. For in 1816 Prout noticed that the atomic weights then known appeared to be integral multiples of that of hydrogen and went on to suggest that the atoms of all elements might be built up out of hydrogen atoms. At a later date, the discovery of fractional atomic weights set Prout's hypothesis in doubt, but the situation changed when results became available from the mass spectrograph. None of those results ever revealed a positively charged particle lighter than the hydrogen nucleus, or proton. With the subsequent formulation by Aston of his whole number rule and the ejection of protons from various nuclei by Rutherford and Chadwick, Prout's hypothesis was immediately revived but was now expressed in terms of atomic nuclei rather than complete atoms. At this time, therefore, an atomic nucleus was believed to contain a number of protons equal to its mass number.

It was now known, however, that the positive charge on the nucleus amounted in magnitude to a number of electronic charges equal to the atomic number, which is only half or less than half the mass number. In order to obtain the correct nuclear charge, therefore, it became necessary to postulate the existence within the nucleus of an appropriate number of electrons. The nucleus thus finally appeared as a compact assemblage of a number of protons equal to the atomic number and a number of proton–electron pairs equal to the difference between the mass number and atomic number.

The discovery of the neutron, however, raised a new issue. On the view of nuclear structure prevailing at the time, the neutron appeared to be compounded from a proton–electron pair such as existed in atomic nuclei. This view found support in the value then adopted for the mass of the neutron which was less than the combined mass of a proton and electron. It followed that the neutron could only be forcibly disrupted into its supposedly constituent proton and electron, if energy were supplied, as indeed it was required to be if the neutron were to have any stability at all. As pointed out by Chadwick in 1933, however, the question then arose as to why the hydrogen atom did not readily transform to a neutron with a release of energy. He was therefore led to suggest that the true mass of the neutron might be very close to the highest value so far put forward and thus slightly exceed the combined mass of a proton and electron. In this event, forcible rupture of a neutron into a proton and electron would involve a release of energy rather than the supply that the evident stability of the neutron demanded. It thus appeared that the neutron was not compounded of a proton and an electron but was a fundamental particle and, close upon its discovery, W. Heisenberg put forward the modern view of nuclear structure in which neutrons replaced the proton–electron pairs postulated hitherto. Since they are the constituent particles of atomic nuclei, protons and neutrons are often collectively referred to as 'nucleons'.

9.3 Binding energy

The closeness of isotopic weights to whole numbers was seen to produce a marked impulse to the thinking on nuclear structure, but their very departures from whole numbers took on a new significance with the confirmation of the Einstein mass–energy relationship. Even before this relationship was generally accepted, Aston had, in 1927, drawn attention to the systematic way in which the fractional deviation of the isotopic weight from the mass number, or 'packing fraction', varied with the mass number. Of greater interest, however, in view of the Einstein relationship, is the amount by which the mass of the nucleus is less than the masses of the several protons and neutrons of which it is composed, for this amount gives at once the energy required to separate the nucleus into its constituent protons and neutrons. This energy, termed the 'binding energy', is clearly a measure of nuclear stability. In the case of a nucleus of mass M amu, mass number A and atomic number Z, the binding energy is

$$B = 931\,[ZM_p + (A-Z)M_n - M] \quad \text{MeV}$$

where M_p, M_n are the respective masses in amu of the proton and neutron.

Like Aston's packing fraction, the binding energy per nucleon varies in a systematic way with the mass number, as shown by the curve of Figure 9.2 for the case of stable isotopes. The ordinate of this curve is the binding energy B divided by the mass number A, and thus gives only the *average* binding energy per nucleon. From the shape of the curve, however, it is clear that as the mass number increases to 55 or so, each added nucleon is bound more and more tightly, the binding energy of those finally added being about 8.75 MeV. With further increase in mass number, additional nucleons are seen to be progressively less tightly bound. Points on the curve removed from the maximum thus correspond to nuclei in which certain nucleons are bound even less tightly to the remainder than the ordinates of these points would imply. This observation is consistent with the limitation of Rutherford and Chadwick's success in artificial transmutation to the very light nuclei and with the instability evidenced in very heavy nuclei by their natural α-decay.

By way of example, the natural α-decay of heavy nuclei may be examined quantitatively from the viewpoint of binding energy. The binding energy of an α-particle, which is released in its formation from two protons and two neutrons, is

$$931\,[2(1.00783 + 1.00867) - 4.00260] = 28.3 \text{ MeV}$$

Of this energy, however, at least 4.0 MeV will supply the α-particle with its kinetic energy and not more than 24.3 MeV will thus be available for detaching its constituent protons and neutrons from the active nucleus. The mass number of this nucleus must then be such that the binding energy per additional nucleon does not exceed

$$24.3/4 = 6.08 \text{ MeV}$$

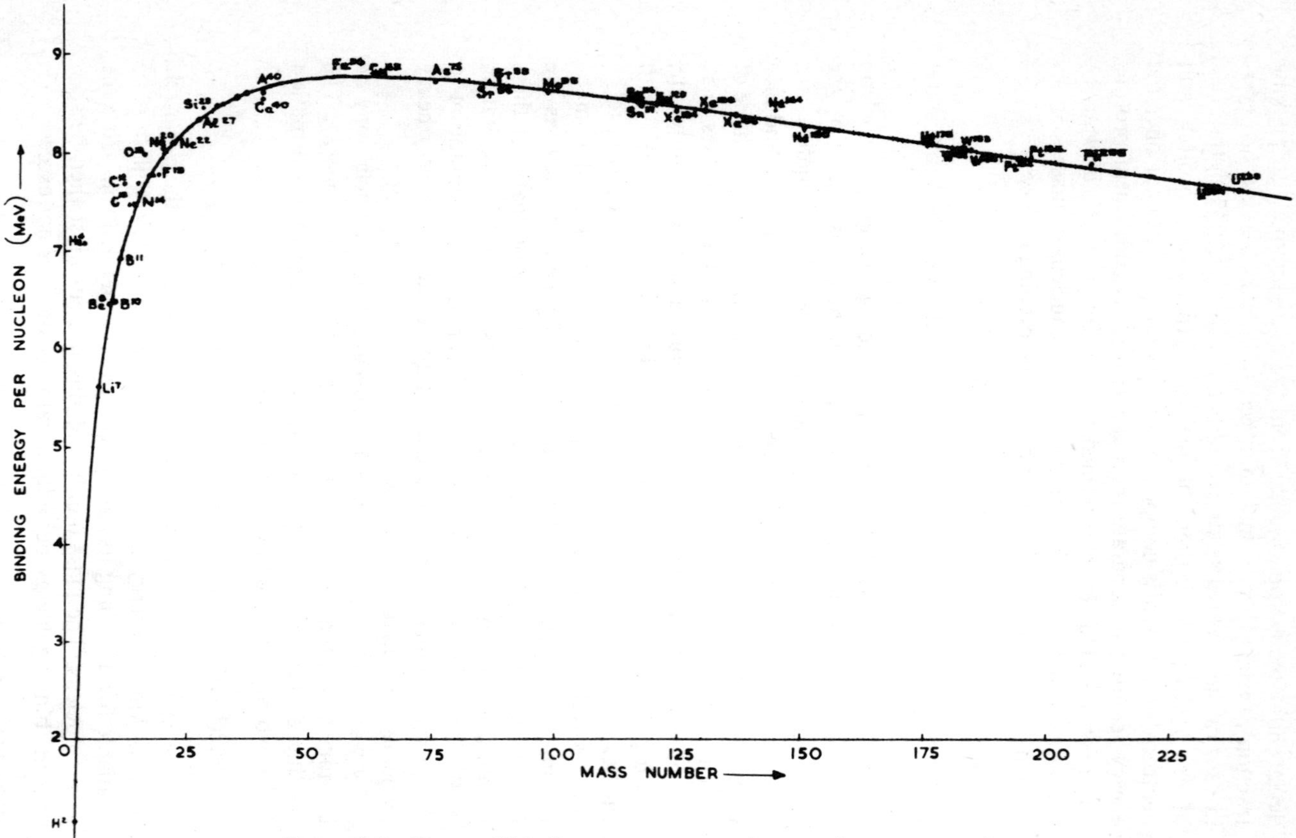

Figure 9.2 Curve of binding energy per nucleon against mass number

At any mass number A, however, the binding energy per additional nucleon is clearly

$$dB/dA = A\frac{d(B/A)}{dA} + (B/A)$$

Also, at mass numbers greater than 150, Figure 9.2 shows that

$$B/A \approx 9.48 - 0.0080A \quad \text{MeV}$$

so that the binding energy per additional nucleon becomes

$$9.48 - 0.0160A \quad \text{MeV}$$

If this expression is not to exceed 6.08 MeV, then A must be greater than 212, which is reasonably close to 210, the lowest mass number of an α-active nucleus in any radioactive series, as seen from Appendix 3.

9.4 Nuclear forces

The experiments described in Section 7.3, which verified Rutherford's calculation of α-particle scattering, showed at the same time that the Coulomb law of electrostatic force persisted to the immediate neighbourhood of the nucleus itself. In the modern view of nuclear structure, this force is attributed to the positive charges carried by the protons within the nucleus. Moreover, if these protons were subject only to the same law of force as applies outside the nucleus, they would mutually repel one another and fly apart. That, far from doing so, they evince a considerable binding energy must therefore be attributed to the existence within the nucleus of attractive forces between the nucleons, sufficiently powerful to provide this energy despite the mutual electrostatic repulsion between the protons. These attractive forces, termed 'nuclear forces', are not yet fully understood, though some conclusions about them may be drawn.

In the first place, an estimate may be made of the energy expended in overcoming the mutual electrostatic repulsion between a nucleus deficient in one proton and the missing proton being brought into it. If, in the notation of Section 7.7, the proton finally occupies a site distant $b/2$ from the centre of the nucleus, the atomic number of which then becomes Z, the energy expended will be

$$\frac{(Z - 1)e^2}{\varkappa b/2}$$

On introducing the same numerical values as in Section 7.7, the energy expended amounts to 4.6 MeV. The average energy expended per nucleon will be less than half this amount. The energy that the nuclear forces must supply to nullify that stored as a result of the mutual electrostatic repulsion between the protons in a nucleus is thus considerably less than that that they must additionally supply in providing the nucleus with its binding energy.

The magnitude of the binding energy per nucleon shows that nuclear forces are of exceptional intensity. If, by way of comparison, the energy required to stretch a piece of steel to its breaking point is divided by the number of nucleons contained in it, the result is about 10^{-4} eV/nucleon. Binding energies per nucleon are thus over 10^{10} times as great. Furthermore, nuclear forces must be of a very short-range character. Otherwise, within a nucleus, they would be exerted not only between adjacent nucleons but between distant ones as well. The binding energy of the nucleus would then vary roughly as the square of the total number of constituent nucleons, whereas it is more nearly directly proportional to this number.

10

Artificial radioactivity

10.1 Cosmic rays

The interest so far shown in the neutron as a constituent of the atom was now overshadowed by that aroused in it as an incident particle. For, being uncharged, it was expected to enter a positively charged nucleus with ease and so more readily promote a nuclear reaction. This expectation was amply fulfilled in the large number of neutron reactions soon afterwards obtained. Furthermore, many of these reactions gave rise to what is known as 'artificial radioactivity', and it was the study of this activity that eventually showed the way to the release of nuclear energy on an industrial scale. Artificial radioactivity had become known shortly before it was first observed in reactions induced by neutrons, and the particle ejected in its first manifestation had been discovered somewhat earlier still. This particle was the 'positron' and its discovery arose from the study of 'cosmic rays', which had by then engaged attention for some thirty years.

In 1900, C. T. R. Wilson concurrently with J. Elster and H. Geitel, had observed the continuous leakage of current from charged electroscopes, despite the provision of shielding against known ionising radiations. A few years later, J. C. McLennan and E. F. Burton found that the leakage was still maintained, albeit at a reduced rate, from an electroscope on the surface of a deep lake, so that only part of the leakage could be attributed to radioactive impurities in the earth. At about the same time, however, Rutherford and H. L. Cooke showed that shielding the electroscope, even with thick layers of iron and lead, only eliminated part of the leakage and scarcely affected the remainder. There was now evidence, therefore, of the universal existence of some ionising radiation of marked penetrating power and not entirely of terrestrial origin.

The study of this radiation acquired renewed interest when, in 1910, A. Gockel took electroscopes up to various heights in balloon ascents. The ionisation was found to diminish as the height increased to about 2000 metres, but beyond that height the ionisation became more intense and, at 5000 metres, the greatest height reached, it exceeded that at the earth's surface. A year later these experiments were repeated by V. F. Hess, using an ionisation chamber in conjunction with an electrometer. By day and by night and also during a solar eclipse he obtained results similar to those of Gockel, and the sun as a source of the ionising radiation was eliminated. By 1914 W. Kolhörster had extended the height of balloon ascent to 9000 metres, and found the ionisation there to be some eight times greater than that at sea level. Unmanned balloons were used by R. A. Millikan in 1922 and, at about the same time, by E. Regener. These burst at pre-arranged heights and recording instruments which they had so far carried then descended to earth by parachute. By this means, heights of 15 500 metres were reached, but the ionisation was found to have reached a maximum at a slightly lower altitude, being there over 100 times as great as that at sea level and on the point of diminishing once more.

It thus appeared that the ionisation was produced by secondary rays arising from interaction between the earth's atmosphere and some form of primary rays which could only be coming from outer space. Clearly, it was these primary rays that exhibited such outstanding penetrating power at the earth's surface, after passing through an atmosphere capable of attenuating γ-rays to the same extent as 760 mm of mercury. In fact, measurements made by Millikan and Regener at great depths in lakes showed that over 200 times the depth of water was required to produce the same attenuation in these primary rays as in the γ-rays from thallium-208. In 1925, Millikan referred to these primary rays as 'cosmic rays' and this is the name by which they are now known.

10.2 The positron

Today the subject of cosmic rays still offers many unsolved problems, but of especial importance for present purposes are the cosmic ray investigations that brought to light new fundamental particles. The first of these particles was discovered by C. D. Anderson in 1932. At the time Anderson was investigating the nature of cosmic rays by means of an apparatus, outlined in Figure 10.1, which had first been used by D. Shobelzyn in 1927 specifically for cosmic ray work. Across

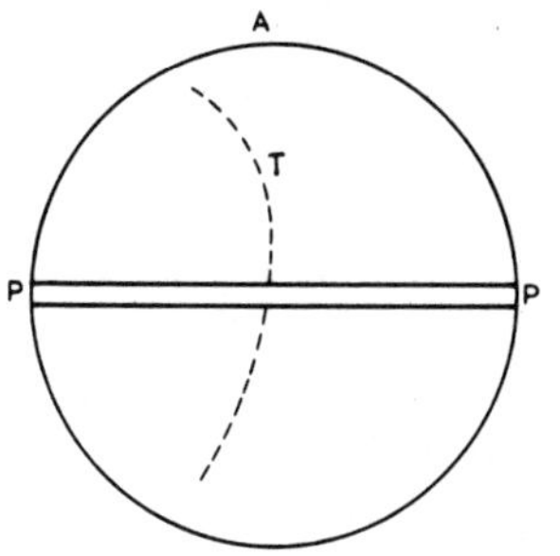

Figure 10.1 Cloud track of positron

the cylinder A of a Wilson cloud chamber was placed a 6 mm thick lead plate PP, so that the cloud space was divided into two semi-cylindrical halves. A strong magnetic field directed along the axis of the chamber caused the cloud track T of any moving charged particle to be curved. If the particle lost energy on passing through the lead plate, it would travel in a path of increased curvature on the far side of the plate and the direction in which it was travelling could thus be inferred.

Since cosmic rays would inevitably be passing through the apparatus from time to time, random exposures with it would reveal an occasional cosmic ray track. Among these tracks, however, Anderson obtained one of the form shown in Figure 10.1. Knowing the direction in which the particle making the track must have been travelling and also that in which the magnetic field acted, the curvature of the track at once gave the sign of the electric charge that the particle carried. This sign was found to be positive. Since the track made by the particle was similar in density to that made by an electron, Anderson concluded that the particle had the mass of an electron and carried an electric charge equal in magnitude to that carried by an electron but of opposite sign. To this particle Anderson gave the name 'positron'. Anderson's conclusion as to the nature of the positron was confirmed in 1934 when J. Thiebaud measured its charge-to-mass ratio by means similar to that used by Thomson in the case of the electron. Thiebaud found that the charge-to-mass ratio for the positron was within 15 per cent of that for the electron.

10.3 **Discovery of artificial radioactivity**

Cosmic rays provided only a randomly intermittent source of positrons, but a continuous source was discovered by F. Joliot and I.

Curie-Joliot. As mentioned in Section 9.1, the Joliots had studied the action of α-particles on light elements. In 1934, on extending this study, they found that a thin foil of aluminium, onto which α-rays had been directed, continued to affect a counter tube even after the α-ray source had been removed. Not only did the counting rate fall off exponentially with time, but by means of an expansion chamber in a magnetic field the ejected particles were found to be positrons. Similar results were obtained with foils of boron and magnesium, and thus it was concluded that certain nuclear reactions yielded residual nuclei in the form of hitherto unknown radio-isotopes. Nuclear transmutations of this type are said to give rise to 'artificial radioactivity'.

For the reaction between α-particles and aluminium, the formula suggested by the Joliots was

$$^{27}_{13}\mathrm{Al} + {}^{4}_{2}\mathrm{He} \rightarrow {}^{30}_{15}\mathrm{P} + {}^{1}_{0}\mathrm{n}$$

$$^{30}_{15}\mathrm{P} \rightarrow {}^{0}_{+1}\mathrm{e} + {}^{30}_{14}\mathrm{Si}$$

where the symbol ${}^{0}_{+1}\mathrm{e}$ denotes the positron. According to this suggestion, the aluminium was transmuted into the radio-isotope phosphorus-30. This isotope then decayed by the emission of a positron to the stable, naturally occurring isotope silicon-30. To confirm their view as to the radio-isotope, the Joliots put it to a chemical test. Aluminium foil that had been exposed to α-particles was dissolved in acid and stable compounds of elements of atomic number close to that of aluminium were then added to the resulting solution. By subsequent chemical separation, the small number of active atoms were deposited in association with a large amount of a stable isotope of the same element. Only deposits containing phosphorus were found to be active, thus proving that the positron emitter was indeed a phosphorus isotope. This procedure gave, for the first time, chemical evidence of artificial transmutation. Furthermore, it enabled the chemical nature of the residual nucleus to be identified directly rather than inferentially as by cloud tracks.

Equations (6) may be applied to cases of artificial radioactivity provided that the rate R_0 of formation of the parent radio-isotope is added to the right-hand side of the first equation. After a period of time that is long in comparison with the longest half-life of the radioactive series, the activities of all isotopes of the series will 'saturate' so that

$$\lambda_0 P_0 = \lambda_1 P_1 = \lambda_2 P_2 = \ldots = \lambda_{s-1} P_{s-1} = R_0$$

10.4 **Artificial radioactivity induced by neutrons**

Within a year of its discovery, artificial radioactivity was becoming increasingly evident, particularly in the case of neutron reactions, on which interest was now strongly centred. Notable among those studying such reactions were E. Fermi and his co-workers E. Amaldi, O. D'Agostino, F. Rasetti and E. Segré, who published their first results in 1934. Their neutron source was a sealed glass tube 6 mm in diameter and 15 mm long, containing beryllium powder and up to 800 millicuries of radon. The samples of the materials tested were generally in the form of cylindrical shells and were irradiated by placing them round the source until any activity had saturated. The samples were then placed round a Geiger–Müller counter 14 mm in diameter and 50 mm long and the half-life of the activity thereby determined. That this activity was in no way due to γ-radiation from the source was proved by showing that radon without beryllium produced no activity in any of the materials tested. Altogether, 64 elements were examined in this way and, of these, 36 were found to be activated by neutrons. In many cases the activated elements were chemically processed and the radio-isotopes thus identified.

The element uranium, however, raised difficulties for the radio-isotopes created in it could not be chemically identified with any of the six elements successively lower in atomic number immediately below that of uranium. Fermi and his colleagues were therefore led to the view that the activity in this case was due to radio-isotopes that had atomic numbers greater than that of uranium and were thus isotopic with 'transuranic' elements. Events were to show that this view did indeed account for a small part of the activity, but the explanation for the remainder was to be far more astounding than either Fermi or his co-workers had suspected.

10.5 **Identification of artificially created radio-isotopes**

Radio-isotopes of all the known elements have now been artificially created by various nuclear reactions and number over seven hundred. A selection of these isotopes is given in Appendix 4.

When artificial radioactivity occurs as the result of a nuclear reaction in an element having several naturally occurring isotopic forms, it is not immediately clear which of these forms has supplied the target nucleus. There is then a need to determine not only the chemical nature of the radio-isotope produced but its mass number as well. This determination is generally made by the 'cross-bombardment method'. A different type of nuclear reaction is used, with a different target element if necessary, to obtain the same radio-isotope, as evidenced by its half-life and a chemical determination of the element with which it is isotopic. The formulae for this and the original reaction may then be applied to each naturally occurring isotope of the appropriate target elements and possible mass numbers of the radio-isotope thus obtained. In general, it will be found that only one of these mass numbers is common to both reactions and this mass number is the one required.

10.6 **Nuclear isomers**

It has so far been assumed that the half-life of a radio-isotope is unique, but in 1917 F. Soddy had suggested the possibility of otherwise identical radio-isotopes exhibiting activities of different half-lives. Four years later O. Hahn found that the naturally occurring radio-isotope protactinium-234 did in fact exist in two states, each decaying with a different half-life. No further cases of such existence were discovered until 1935, when they began to appear in several quarters as a result of the growing number of radio-isotopes artificially created. In the following year Lise Meitner introduced the term 'nuclear isomers' to denote the several states of a radio-isotope characterised by different rates of decay. The decay of a radio-isotope from one such state to another is referred to as 'isomeric transition'. At the present time, over seventy cases of isomeric transition are known. In most of these, the transition is only from one isomer to another, but in a few instances two successive transitions occur.

11

Neutron collisions

11.1 Cross-section

Since the neutron is uncharged, its interaction with an atomic nuleus will be negligible until the two are very close together and the interaction itself may then be expected to take the form of an elastic collision. In fact, it was on the hypothesis of just such a collision that Chadwick adduced evidence for the existence of the neutron and for the magnitude of its mass. Other forms of collision may, however, occur, such as those resulting in activation of the struck nucleus. Furthermore, a neutron may collide with nuclei of more than one type, unless the material in which it is travelling is a monisotopic element. As to whether a collision of particular form is in fact made with a nucleus of given type is a question of probability. The probability is that of a neutron making the particular form of collision with a nucleus of the given type in unit distance of travel. On the assumption that the probability is unaffected by the distance travelled it may be expressed by a quantity Σ independent of this distance.

It is evident that the probability is proportional to the number of nuclei of the given type in unit volume of the material through which the neutron is travelling. If, in unit volume of the material, ϱ is the mass of all the nuclei of the given type, then for collisions of the particular form with nuclei of this type,

$$\Sigma = \sigma N_0 \varrho / \mathscr{A} \qquad (18)$$

where σ is a constant of proportionality depending only on the velocities of the colliding neutron and nucleus collided with, the form of the collision and the type of nucleus collided with.

The quantity σ may be given a physical significance. Referring to

Figure 11.1, a right prism of the material has a cross-sectional area, a_R, normal to the direction of travel of a neutron, and a length in this direction equal to l. With each nucleus of the given type is associated a target area a normal to the direction of neutron travel. The sum of such target areas is then

$$a \frac{N_0 \varrho}{\mathscr{A}} a_R l$$

so that the probability of the neutron passing through one of them is

$$a \frac{N_0 \varrho}{\mathscr{A}} l$$

The corresponding probability per unit distance of neutron travel is therefore

$$a N_0 \varrho / \mathscr{A}$$

On comparing this expression with (18), it is seen that σ is the target area associated with a nucleus of the given type, such that, in unit distance of travel, the probability of a neutron traversing a normal to the area is the same as that of it meeting the nucleus in a collision of the particular form. The quantity σ is thus assigned the name 'cross-section'. The same name is also given to the probability Σ and, to distinguish between the two cross-sections, that denoted by σ is termed 'microscopic' and that by Σ 'macroscopic'. In what follows the term 'cross-section' will be understood to refer to the macroscopic variety Σ unless otherwise stated.

The microscopic cross-section σ is completely independent of the way in which the nuclei of the given type are disposed in a material, whether they belong to atoms combined with other atoms as a molecule of a compound, to atoms of one of a mixture of elements or to atoms of which the material is entirely composed. On the other hand, the mass ϱ of nuclei of the given type per unit volume of material will vary with the way in which they are disposed in it and so, therefore, will the macroscopic cross-section Σ. The value of Σ appropriate to any such disposition is thus obtained by multiplying $N_0\sigma/\mathscr{A}$ by the appropriate value of ϱ. If the nuclei of given type form a proportion of weight w of a material of density ϱ_R, then the appropriate value of ϱ is merely $w\varrho_R$. In the case of a material formed of several types of atom denoted by suffixes A, B, ..., therefore, the macroscopic cross-sections for collisions with nuclei of types A, B, ... are respectively

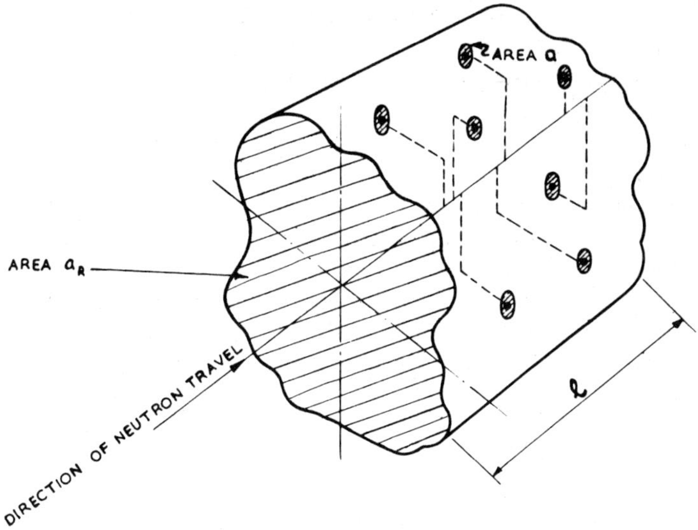

Figure 11.1 The nuclear cross-section

$$\Sigma_{A/R} = \varrho_R N_0 w_A (\sigma/\mathscr{A})_A$$
$$\Sigma_{B/R} = \varrho_R N_0 w_B (\sigma/\mathscr{A})_B \text{ etc.}$$

The resultant cross-section is then

$$\Sigma_R = \Sigma_{A/R} + \Sigma_{B/R} + \ldots$$

When the material is simply a mixture of elements, ϱ is the density of the pure element times its fraction by volume in the material. The expression for the macroscopic cross-section of the material thus becomes

$$\Sigma_R = x_A \Sigma_A + x_B \Sigma_B + \ldots$$

where x_A, x_B, ..., denote the fractions by volume of the elements A, B, ..., in the material, and Σ_A, Σ_B, ... the cross-sections of the pure elements A, B, ...

The macroscopic cross-section is measured in units of (1/length) and is a more useful concept when dealing with the behaviour of neutrons in a material. The microscopic cross-section is measured in units of area. The customary unit is 10^{-24} cm^2, known as a 'barn' and denoted by the symbol b. This sort of cross-section finds its greatest value in a study of the effects of atomic constituents and of neutron and nuclear velocities.

If σ_α is the microscopic cross-section of an atom for one form of collision, σ_β that for another form and so on, any of which may occur independently, then the resultant cross-section of the atom for such forms of collision is

$$\sigma_r = \sigma_\alpha + \sigma_\beta + \ldots$$

When the cross-sections σ_α, σ_β, ... extend to all possible forms of collision, any of which may occur independently, the resultant cross-section is known as the 'total' cross-section, σ_t. In general, a cross-section must be given suffixes to denote both the atom and the forms of collision to which it applies, so that the cross-section of an atom N for collisions of the forms r would be written σ_{r_N}.

11.2 Collision energy

Among the factors governing the cross-section will be the velocities of the neutron and nucleus collided with. It is reasonable to assume, however, that the cross-section is unaffected by the imposition of a uniform velocity on the medium in which the nucleus is contained so that it is the relative velocity of colliding neutron to nucleus that is significant rather than the velocity of either. If neutron and nucleus have velocities before collision denoted by V and U respectively, then their relative velocity V_r may be found from a velocity triangle as shown in Figure 11.2.

Further discussion will make use of the concept of the 'centre-of-mass' of neutron and nucleus. This must clearly lie on the straight line joining the two, and be such that its distances from nucleus and neutron are in the ratio of the mass M_n of the neutron to the mass of the nucleus, which is closely approached by AM_n or, more accurately, for mass numbers over 10, by 1 per cent less than AM_n. Since the line joining neutron and nucleus will be extending at the rate V_r, it follows that the velocity U_{c_r} of the nucleus relative to the centre-of-mass and that V_{c_r} of the neutron will be in the same ratio, so that

$$U_{c_r}/V_{c_r} = M_n/AM_n$$

or

$$V_{c_r} = AU_{c_r} \tag{19}$$

Referring to Figure 11.2, the velocity V_c of the centre-of-mass itself is defined by the line joining the apex formed by V and U to the point in V_r which subdivides it into V_{c_r} and U_{c_r}, making an angle γ with V_r.

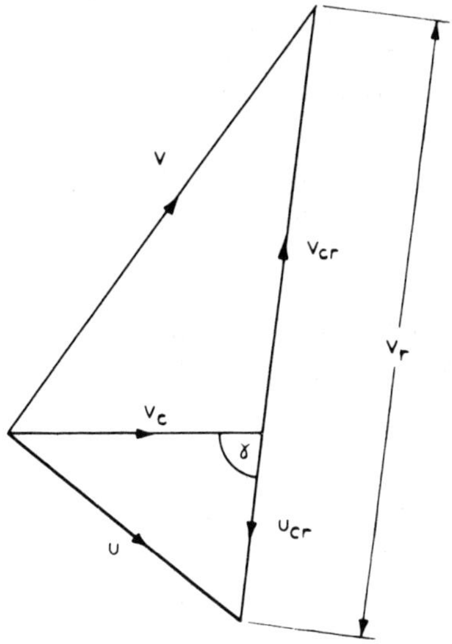

Figure 11.2 Velocities of neutron and nucleus before collision

The energies of neutron and nucleus before collision may be respectively written as

$$\tfrac{1}{2} M_n V^2 = \tfrac{1}{2} M_n[V_c^2 + V_{c_r}^2 + 2V_c V_{c_r} \cos \gamma]$$

$$\tfrac{1}{2} AM_n U^2 = \tfrac{1}{2} M_n[A V_c^2 + AU_{c_r}^2 - 2A V_c U_{c_r} \cos \gamma]$$

On adding these two equations and introducing (19)

$$\begin{aligned} &\tfrac{1}{2} M_n V^2 + \tfrac{1}{2} AM_n U^2 - \tfrac{1}{2}(A + 1) M_n V_c^2 \\ &= \tfrac{1}{2} M_n \left[\left(\frac{A}{A+1}\right)^2 + A\left(\frac{1}{A+1}\right)^2\right](V_{c_r} + U_{c_r})^2 \\ &= \tfrac{1}{2} M_n \frac{A}{A+1} V_r^2 \end{aligned} \tag{20}$$

The expression on the right-hand side of this equation is known as the 'collision energy' ϵ. Since it is a function of no velocity other than V_r, it is a quantity in terms of which the velocity dependence of cross-section may strictly be expressed. The quantity $[A/(A+1)]M_n$ is referred to as the 'reduced mass'. Except for the lightest nuclei, the reduced mass is very nearly the mass of the neutron. Also, U is commonly small compared with V so that V_r approaches V. The collision energy $\in$ may then be generally identified with the neutron energy E. Furthermore, the cross-section is frequently required as a function of E and it is therefore usually expressed in terms of E rather than $\in$.

The cross-section of an atom N for collisions of the forms r at a neutron energy E_i will be written

$$(\sigma_{rN})_{E_i}$$

In many cases the cross-section will apply to neutrons of a particular energy distribution i, and will then be written

$$\sigma_{rNi}$$

If the reference to the cross-section is irrespective of the type of atom or forms of collision, the suffixes N and r may be omitted. In the same way, the suffix E_i or i may be omitted if the reference is irrespective of the neutron energy or energy distribution.

11.3 Compound nucleus

From equation (20) the collision energy is seen to be the decrease in kinetic energy of neutron and nucleus if they coalesce on collision. In this event the collision energy represents the energy available for producing any consequent nuclear reaction. A suggestion on these lines was in fact put forward by N. H. D. Bohr in 1936.

From the half-lives of nuclei activated by neutrons Bohr concluded that an encounter between a high-speed neutron and a heavy nucleus lasted much longer than the time taken by the neutron to pass through a space such as the nucleus occupied. He was thus forced to assume that the first stage of a collision between a high-speed neutron and a heavy nucleus was their coalescence into a 'compound nucleus' of remarkable stability. Subsequently, the compound nucleus might undergo radioactive decay or, in the case of a scattering process, emit a further neutron. These possibilities however were to be considered as separate competing events having no immediate connexion with the first stage of the collision. On formation of the compound nucleus, the excess energy of the colliding neutron would be rapidly shared between all the nuclear particles and for some time afterwards no single particle would possess sufficient kinetic energy to leave the nucleus. The subsequent liberation of a particle thus implied a complicated process in which sufficient of the excess energy became concentrated on the particle at the surface of the nucleus.

11.4 Mean free path

In travelling a distance ds, a neutron has a probability of making a collision equal to

$$\Sigma \, \mathrm{d}s$$

where Σ is the cross-section for the particular form of collision with the given type of nucleus. The probability of the neutron escaping such collision is thus

$$1 - \Sigma \, \mathrm{d}s$$

In travelling a second successive distance ds, the probability of the neutron still escaping such a collision is

$$(1 - \Sigma \, \mathrm{d}s)^2$$

and in travelling a distance s this probability becomes

$$(1 - \Sigma \, \mathrm{d}s)^{s/\mathrm{d}s} = 1 - s\Sigma + \frac{s^2}{2!}\left(1 - \frac{\mathrm{d}s}{s}\right)\Sigma^2 - \frac{s^3}{3!}\left(1 - \frac{\mathrm{d}s}{s}\right)\left(1 - 2\frac{\mathrm{d}s}{s}\right)\Sigma^3 + \ldots$$

on expanding by the binomial theorem. Since ds/s is vanishingly small, the probability of the neutron escaping collision in the distance s is

$$1 - \Sigma s + \frac{1}{2!}(\Sigma s)^2 - \frac{1}{3!}(\Sigma s)^3 + \ldots = \mathrm{e}^{-\Sigma s} \qquad (21)$$

Expression (21) gives the proportion of neutrons that have travelled a distance s without collision of the particular form with a given type of nucleus. The proportion that make such a collision in the next successive increment of distance ds is therefore

$$\mathrm{e}^{-\Sigma s} \Sigma \, \mathrm{d}s$$

and this is also the proportion that travel a distance varying from s to s + ds between successive collisions. The average distance travelled by a neutron between successive collisions is therefore

$$\begin{aligned}\lambda &= \int_0^\infty \Sigma s e^{-\Sigma s}\, ds \\ &= -\frac{1}{\Sigma}\left[(\Sigma s + 1)\, e^{-\Sigma s}\right]_0^\infty \\ &= 1/\Sigma\end{aligned}$$

and is known as the ‘mean free path’.

12

Fast neutrons

12.1 Total cross-sections for fast neutrons

Neutrons that have travelled a distance less than their total mean free path will have made relatively few collisions with atomic nuclei and so have energies approaching those at which they were emitted by the source. They will therefore be referred to as 'fast' neutrons. The energies of neutrons emitted by a radon–beryllium source were determined by J. R. Dunning in 1934 with the apparatus shown diagrammatically in Figure 12.1. The source S consisted of a glass bulb packed with fine beryllium filings and filled with radon. The bulb was housed in a thick platinum container and supported on a long thin rod. A lead cylinder G, 30 mm thick, served as a gamma ray shield. Neutrons were detected by a parallel plate ionisation chamber I having a paraffin disc P, 2 mm thick, immediately in front of it. The pulses recorded by the chamber were due partly to protons knocked on from the paraffin disc and partly to oxygen and nitrogen nuclei projected from the chamber air. Material in the form of a cylindrical sample M could be placed between the chamber and the source.

In the absence of the cylindrical sample, the counts with the paraffin removed were subtracted from those with it in place to give the counts due to protons knocked on from it. By inserting various thicknesses of mica between the paraffin and the ionisation chamber, the relative numbers of protons, and therewith neutrons, above

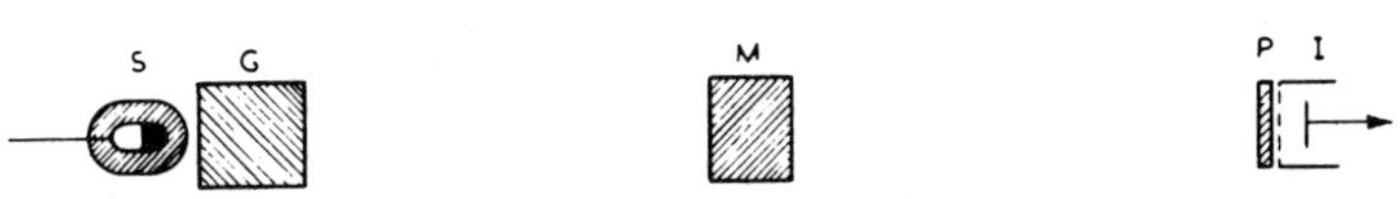

Figure 12.1 Production, scattering and detection of fast neutrons

different maximum velocities could be obtained. In this way the neutron energy was found to vary from a maximum of 14.2 MeV down to about 0.5 MeV, a comparatively large proportion of neutrons having energies in the region of 5.0, 6.5 and 8.0 MeV.

With a cylindrical sample placed between chamber and source, it is clear from Figure 12.2 that, of the neutrons entering the sample, those reaching the chamber were

the proportion which avoided a collision (type A in Figure 12.2)
+ a proportion scattered within the unscattered beam (type B)
+ a proportion scattered outside the unscattered beam (type C)
+ a proportion undergoing multiple scattering

From equation (21), the proportion of type A is clearly

$$e^{-\Sigma_t T}$$

where T is the thickness of the sample and Σ_t its total cross-section. The proportion of type B must be scattered through an angle less than some limiting value determined by the geometrical arrangement of the apparatus. The arrangement was such that the limiting value was 15° and the proportion of type B therefore small. Albeit, on the understanding that the total cross-section Σ_t took no account of collisions in which neutrons were scattered through angles less than this limiting value, the above expression for the proportion of type A gave the proportions of types A and B combined. The proportion of type C was eliminated by making the cylindrical sample fit exactly within the unscattered beam. The proportion undergoing multiple scattering was minimised by so adjusting the thickness of the sample that the proportion entering the chamber was about 40 per cent. In this way, the thickness was prevented from exceeding one total mean free path. Finally, therefore, the total cross-section for all forms of collision, other than those in which neutrons were scattered through angles less than the limiting value, could be found from the relationship

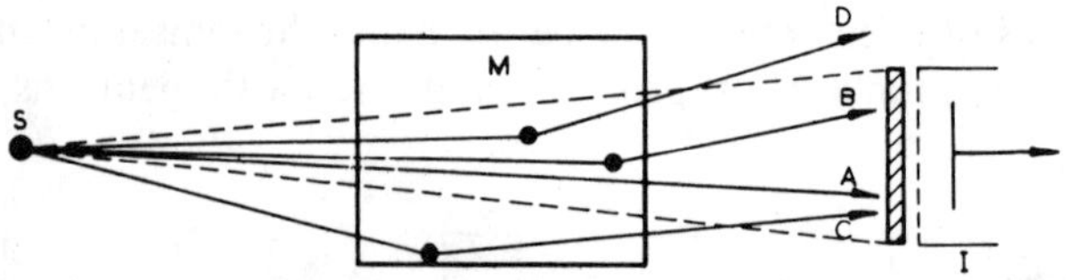

Figure 12.2 Effect of neutron scattering angle on detection

$$\Sigma_t = \frac{1}{T} \log \left(\frac{\text{counts without sample}}{\text{counts with sample}} \right)$$

Fourteen elements were examined in these experiments which, though somewhat primitive by modern standards, gave total cross-sections close to present-day values, even though the precise neutron energy to which the cross-sections applied could not be specified. More recently, the total cross-sections for most elements have been determined as functions of fast neutron energy. These determinations have been made by means of one of the known nuclear reactions in which monoenergetic neutrons are emitted. One such reaction is the ^{7}Li (p,n) ^{7}Be reaction. In this case the target material is bombarded by incident protons that have been artificially accelerated, by a cyclotron for instance, to an energy that can be varied in a controlled manner. In this way, the ejected neutrons may be given any prescribed energy from 0.05 to 3 MeV and used as the neutron source in an apparatus for cross-section measurements.

At 5 MeV the microscopic total cross-section is

$$(\sigma_t)_{5\ \text{MeV}} \approx 0.66\ Z^{0.5} \quad \text{barns}$$

This relation is in error by less than 16 per cent for 27 of the more common elements in the range of Z from 5 to 80.

If, as in Section 7.7, b is taken to be the diameter of a nucleus, and if, in equation (9), all quantities except Z have the same numerical values as in Section 7.7, then

$$(\sigma_t)_{5\ \text{MeV}} \approx 220\ Z^{-1.5}\ (\pi b^2/4) \quad \text{cm}^2$$

The microscopic total cross-section may be identified with a nuclear target area so that, except for the lightest atoms, this area is seen to be of the same order of magnitude as the cross-sectional area of the nucleus itself, an agreement that confirms that reactions between fast neutrons and atomic nuclei are veritable collisions.

12.2 **Fast neutron absorption**

Although collisions between fast neutrons and atomic nuclei were generally believed to result in scattering, the work done by Fermi and his colleagues on activation by neutrons had shown that another form of collision occurred in which the impinging fast neutron was absorbed by the struck nucleus. The relative frequency with which these two forms of collision occurred was investigated in 1935 by J. R. Dunning, G. B. Pegram, G. A. Fink and D. P. Mitchell, and their apparatus is outlined in Figure 12.3. A radon–beryllium neutron source S in a platinum container was located at the centre of a spherical copper vessel V. This vessel was 260 mm in diameter and contained the material under investigation. An ionisation chamber I with a paraffin front enabled neutrons to be detected by protons knocked on from the paraffin and by oxygen and nitrogen nuclei projected from the chamber gas. A 20 mm thickness of lead G constituted a gamma ray shield.

Since the material investigated was disposed symmetrically about the source, neutron scattering within this material did not affect the chamber counts. The fractional reduction in number of counts on placing the material round the source thus gave the proportion of neutrons that were either absorbed by the material or reduced by it to energies less than that required to project protons or chamber gas nuclei. This proportion was then divided by the minimum number of collisions a neutron could make on passing through the material to give the maximum possible proportion absorbed per collision. The number of collisions used in this calculation was obtained by dividing the radius of the copper vessel by the total mean free path, determined as described in the previous section. Tests made in this way on six different materials indicated that not more than 10 per cent of the collisions made by fast neutrons resulted in absorption.

12.3 Elastic scattering

From the fast neutron absorption measurements made by Dunning and his co-workers, most fast neutron collisions were seen to result in scattering, and the elastic nature of this scattering found support from Chadwick's deduction of the existence and mass of the neutron. The

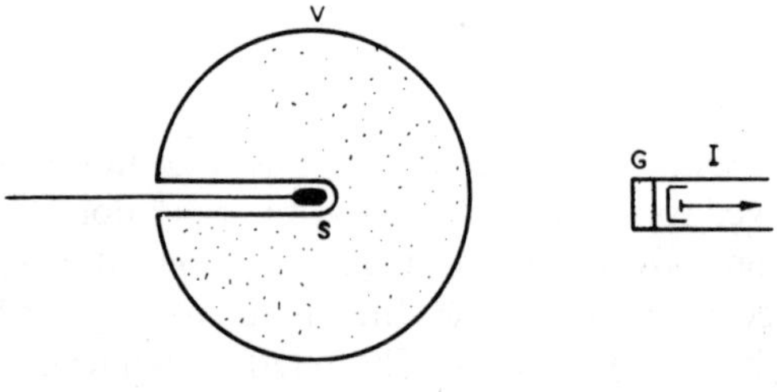

Figure 12.3 Measurement of fast neutron absorption

theory of elastic scattering is thus of importance in studying the behaviour of neutrons in their passage through a material.

If V and M denote respectively the velocity and mass of the neutron, its elastic scattering by a nucleus initially at rest will be governed by equations (13), in which the ratio m/M may now be closely approximated by the mass number A of the scattering nucleus as in Section 11.2. The first two of equations (13) may then be written

$$\left(\frac{v}{V}\cos\psi - 1\right)^2 = A^2\left(\frac{u}{V}\right)^2 \cos^2\varphi$$

$$\left(\frac{v}{V}\sin\psi\right)^2 = A^2\left(\frac{u}{V}\right)^2 \sin^2\varphi$$

and added together to eliminate φ. From the result so obtained u/V may be eliminated by means of the third of equations (13), whence

$$\frac{v}{V} = \frac{\cos\psi + \sqrt{(A^2 - \sin^2\psi)}}{A + 1}$$

The ratio of the energy E of the neutron after collision to that E' before collision is simply the square of v/V so that, finally

$$E/E' = \frac{[\cos\psi + \sqrt{(A^2 - \sin^2\psi)}]^2}{(A + 1)^2} \tag{22}$$

The energy loss of a neutron by elastic scattering is seen to depend partly on the mass number of the scattering nucleus and partly on the angle ψ through which the neutron is scattered. This angle, however, cannot be given a precise value but only a probability of its attainment. Accordingly, in Figure 12.4, the scattering nucleus N is regarded as located at the centre of a sphere of unit radius. A neutron n is supposed to be scattered through unit area of the surface of the

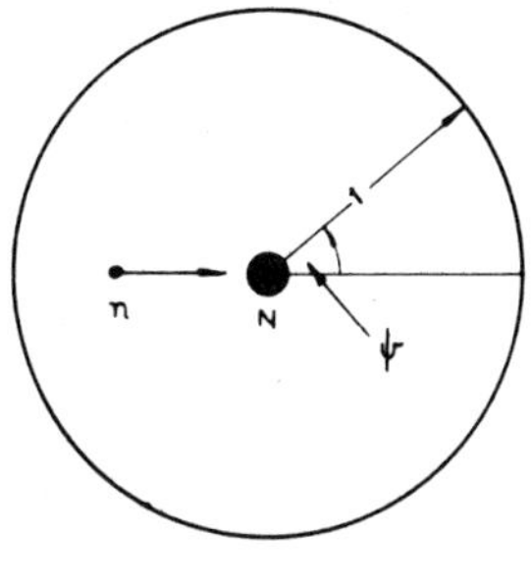

Figure 12.4 Probability of scattering through angle ψ

sphere, or into unit 'solid angle'. If the normal to the centre of this area is inclined to the original line of travel of the neutron at the angle ψ, then the neutron is said to be deflected into unit solid angle through the angle ψ. The probability that a scattered neutron will be deflected in this way is written $p(\psi)$. The probability of the neutron being deflected through an angle between ψ and $\psi + d\psi$ is then

$$p(\psi) \,.\, 2\pi \sin \psi \, d\psi$$

The corresponding probability of the neutron being degraded in energy to an amount between E and $E + dE$ is

$$-2\pi \, p(\psi) \sin \psi \frac{d\psi}{dE} dE \tag{23}$$

which, with equation (22), becomes

$$\frac{\pi}{E'} (A + 1)^2 p(\psi) \frac{\sqrt{(A^2 - \sin^2 \psi)}}{[\cos \psi + \sqrt{(A^2 - \sin^2 \psi)}]^2} dE \tag{24}$$

A further substitution of equation (22) into this expression would give it in terms of E instead of ψ.

This somewhat complicated relation may be replaced by a much simpler one if the angle through which the neutron is deflected is defined as that between its initial direction and its final direction, not in the absolute sense but relative to the centre of mass of neutron and nucleus. Scattering defined in this way is said to be in the 'centre-of-mass' system, that as previously defined being in the 'laboratory' system. The centre-of-mass system may otherwise be defined as one in which scattering is viewed relative to the compound nucleus. On this view elastic scattering is a reaction in which the compound nucleus disintegrates by emitting a neutron in such a way that the total kinetic energy of the emitted neutron and residual nucleus is equal to that of the incident neutron and target nucleus.

Figure 12.5 has been constructed on the lines of Figure 11.2 but the velocities are now those after collision instead of before collision as previously. In order to conserve momentum, the velocity V_c of the centre-of-mass will have the direction of the initial neutron velocity V and a magnitude given by the relation

$$V_c = V/(A + 1) \tag{25}$$

appropriate to a target nucleus at rest. The velocity of the scattering nucleus relative to the centre-of-mass is denoted by u_r which, by analogy with Figure 11.2, will be oppositely directed to the relative velocity v_r of the scattered neutron.

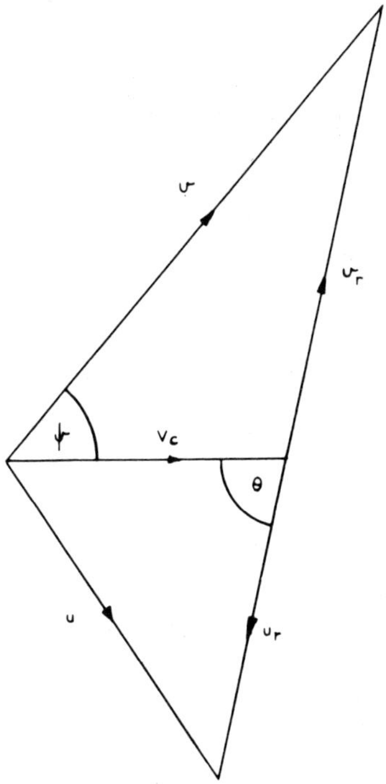

Figure 12.5 Scattering in the centre-of-mass system

Also, analogous to equation (19),

$$v_r = Au_r \qquad (26)$$

Denoting the angle through which the neutron is deflected in the centre-of-mass system by θ, conservation of energy leads to the result

$$V_c^2 + v_r^2 + 2V_c v_r \cos\theta + A(V_c^2 + u_r^2 - 2V_c u_r \cos\theta) = V^2$$

Eliminating v_r by means of equation (26) and V by means of (25), this relation reduces to

$$u_r = V_c$$

whence, from (26),

$$v_r = AV_c \qquad (27)$$

so that

$$v^2 = \frac{V^2}{(A + 1)^2}(1 + A^2 + 2A\cos\theta)$$

Finally, therefore,

$$\begin{aligned} E/E' &= v^2/V^2 \\ &= (1 + A^2 + 2A\cos\theta)/(A + 1)^2 \end{aligned} \tag{28}$$

The probability of the scattered neutron being degraded to energy range dE at energy E is now, analogous to expression (23),

$$-2\pi \mathrm{p}(\theta) \sin\theta \frac{\mathrm{d}\theta}{\mathrm{d}E} \mathrm{d}E$$

where p(θ) is the probability that the neutron is deflected into unit solid angle through the angle θ. From equation (28) this expression becomes, analogous to (24),

$$\frac{\pi}{E'}(A + 1)^2 \frac{\mathrm{p}(\theta)}{A} \mathrm{d}E \tag{29}$$

That the probability of being degraded into a range of energy dE should be proportional not only to the range itself but also to some function of θ, and thus to some function of the energy level E at which the range is situated, may be expected to make further development of the theory of elastic scattering extremely complicated. It is therefore worth examining the case in which p(θ) is so adjusted that the probability expression (29) becomes independent of θ. In this case, it is clear that

$$\mathrm{p}(\theta) = k$$

where k is a constant, so that the probability of degradation into energy range dE becomes independent of the energy level E of the range when scattering is isotropic in the centre-of-mass system. This form of scattering will be seen in the next section to be reproduced fairly well by the lighter atoms, which are important as giving a relatively large reduction in energy per collision. Since

$$2\pi \int_0^\pi \mathrm{p}(\theta) \sin\theta \, \mathrm{d}\theta = 1,$$

it follows that

$$k = 1/4\pi \tag{30}$$

From equation (28) it is clear that the greatest reduction in neutron energy occurs when the angle θ is equal to π, and that in this case (28) reduces to

$$\alpha = [(A - 1)/(A + 1)]^2 \tag{31}$$

where α is the smallest fraction of its incident energy to which the neutron can be reduced in a single scattering collision. On introducing equations (30) and (31) into expression (29), the probability of the neutron being degraded into the energy range $\mathrm{d}E$ becomes, finally,

$$\mathrm{d}E/[(1 - \alpha)E'] \tag{32}$$

The centre-of-mass system may be converted to the laboratory system in the following way. From Figure 12.5,

$$v \cos \psi = V_c + v_r \cos \theta$$

This equation becomes, on introducing (27), (25) and (28),

$$\cos \psi = (1 + A \cos \theta)/\sqrt{(1 + A^2 + 2A \cos \theta)} \tag{33}$$

Also,

$$\mathrm{p}(\psi) \sin \psi \, \mathrm{d}\psi = \mathrm{p}(\theta) \sin \theta \frac{\mathrm{d}\theta}{\mathrm{d}\psi} \mathrm{d}\psi \tag{34}$$

whence

$$\begin{aligned} \mathrm{p}(\psi) &= \frac{\mathrm{p}(\theta)}{\mathrm{d}(\cos \psi)/\mathrm{d}(\cos \theta)} \\ &= \mathrm{p}(\theta)(1 + A^2 + 2A \cos \theta)^{3/2}/A^2(A + \cos \theta) \end{aligned} \tag{35}$$

from equation (33). Given $\mathrm{p}(\theta)$ as a function of θ, therefore, equations (33) and (35) enable $\mathrm{p}(\psi)$ to be expressed as a function of ψ. When A is very large, $\cos \psi$ is seen to be very nearly equal to $\cos \theta$ and $\mathrm{p}(\psi)$ to $\mathrm{p}(\theta)$. At the other extreme, when $A = 1$,

$$\left.\begin{aligned} &\cos \psi = \sqrt{[\tfrac{1}{2}(1 + \cos \theta)]} \\ &\text{so that} \\ &\mathrm{p}(\psi) = 4\,\mathrm{p}(\theta) \cos \psi \end{aligned}\right\} \tag{36}$$

and is zero when ψ exceeds $\pi/2$.

12.4 Scattering angle distribution

In order to apply the results of the previous section, the probabilities $\mathrm{p}(\psi)$ or $\mathrm{p}(\theta)$ must first be determined from measured distributions of scattering angle. One of the earliest experiments on the distribution of scattering angle was made by Dunning in 1934 with the apparatus shown diagrammatically in Figure 12.6. The source S was distant 230 mm from the ionisation chamber I. The direct beam of neutrons from

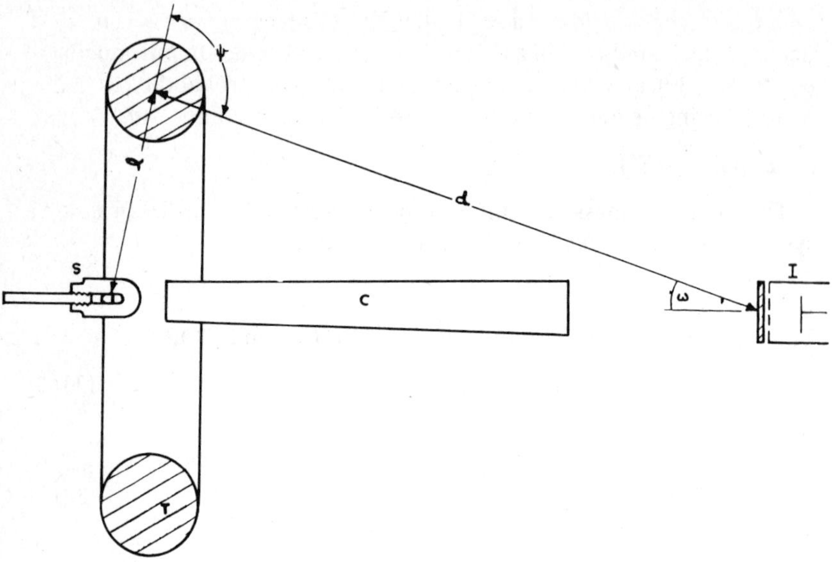

Figure 12.6 Determination of scattering angle distribution

the source was largely scattered away from the chamber by a tapered cylinder C made of platinum which, due to its high total cross-section, is a very effective scattering medium. The distribution of scattering angle ψ was measured by the increase in counts when a toroid T was placed coaxially with source and chamber at various positions relative to them. The mean diameter of the toroid was 134 mm and the diameter of its circular section 36 mm, whilst it was available in different materials. Denoting by l and d the distances from the mean circumference of the toroid to the source and chamber respectively, the recorded increase in the number of counts on introducing the toroid was multiplied by l^2 to allow for the reduced number of neutrons reaching the toroid from the source at increasing distances between them. A further multiplication by d^2 allowed for the reduced number of neutrons reaching the ionisation chamber from the toroid at increasing distances between them. Denoting by ω the angle between the axis of the toroid and a line joining its mean circumference to the chamber, the result thus obtained was finally divided by $\cos\omega$ to allow for the reduced area of the chamber window presented to neutrons reaching it from the toroid with increasing values of ω. Increases in the number of counts, corrected in this way, on introducing toroids of paraffin, carbon, copper and lead were obtained for angles ψ equal to 59° ± 15°, 77° ± 20°, 95° ± 20° and 132° ± 13°.

If A is unity, then equation (36) shows that p(ψ) must be zero when ψ is equal to 90° or more. This value of A corresponds almost exactly to the case of scattering by hydrogen so that Dunning's results for paraffin can only include appreciable scattering by hydrogen at the two smaller values of ψ, scattering at the other two values being almost entirely by carbon. The results obtained for all four materials are consistent with such of the more recent determinations as apply to neutrons of energy from 1.0 to 2.0 MeV. Some of these determinations have been made with a more refined version of the toroid technique used by Dunning. Others have been made with the sample of material to be investigated in the form of a cylinder, the neutron source and detector lying in the same plane at right angles to the sample axis. The detector may then be rotated about this axis to obtain different values of ψ, being screened against direct radiation from the source by a suitable shield. Use of a monoenergetic neutron source enables determinations to be made at different discrete neutron energies.

Results on the distribution of scattering angle are usually presented as a plot of σ_sp(θ) against cos (θ), where σ_s is the scattering cross-section and where cos θ is unity at the origin. The product σ_sp(θ) is written σ_θ, and is the cross-section for scattering into unit solid angle through the angle θ, a quantity known as a 'differential scattering cross-section'. Likewise, the product σ_sp(ψ) is written as the differential scattering cross-section σ_ψ. Substantially isotropic scattering is maintained in the centre-of-mass system up to 15 MeV by hydrogen, up to 1.0 MeV or so by the lighter elements and up to about 0.1 MeV by the heavier elements. Anisotropy generally first appears as a dip in the curve of $(\sigma_\theta)_E$ against cos θ. At higher energies still, $(\sigma_\theta)_E$ falls rapidly as cos θ decreases from 1.0 and then passes through one or more peaks at comparatively small and generally decreasing values.

In addition to elastic scattering as considered here another kind of scattering should be noted. This is 'inelastic' scattering in which the momentum of the colliding neutron and nucleus is conserved but not their kinetic energy, the difference appearing in the form of gamma radiation. Since inelastic scattering only takes place at neutron energies exceeding 0.1 MeV, its significance disappears after only a few scattering collisions and it will not be further considered.

13

Slow neutrons

13.1 Slowing down of neutrons

In 1934, arising from the investigations into neutron-induced activity described in Section 10.4, Fermi and his co-workers found the activation of silver to be strongly influenced by objects surrounding the neutron source. In particular, the activation could be increased to a marked degree by surrounding both source and activated material with a large volume of water or paraffin wax. When the surrounding substance did not contain hydrogen, no comparable increase in activation was observed. That this increase was due to neutrons and not to γ-radiation was confirmed on replacing the radon–beryllium source by radium, which then failed to give any activation at all.

As shown by equation (31), a neutron making a collision in hydrogen may lose its energy completely so that the average reduction in its energy per collision may be expected to approach a half. Little more than twenty collisions would then be needed to reduce its energy from that of the source to that of the surrounding atoms or molecules in thermal equilibrium. Thenceforward any further collisions made by the neutron would as frequently cause an increase in its energy as a similar decrease so that its average energy would remain constant at the thermal equilibrium value. It was therefore assumed that the increase in activation on introducing a hydrogeneous substance was due to the relative ease with which neutrons could be scattered and slowed down in such a substance, whereafter they were further assumed to be more readily captured than fast neutrons. These assumptions were soon to be amply justified.

13.2 Slow neutron capture

Neutrons that have been so slowed down that the activity they induce is enhanced are known as 'slow' neutrons. Capture of a slow neutron, however, does not necessarily result in activation of the capturing nucleus. In particular, slow neutron capture by certain light nuclei gives rise to the emission of α-particles, as was shown by Chadwick in collaboration with M. Goldhaber at the end of the year 1934. The element investigated was enclosed as a target or a gas in an ionisation chamber connected to an amplifier and oscillograph. On exposing the chamber to neutrons from a radon–beryllium source there was always observed a small number of counts due mainly to recoil particles. The source and chamber were then surrounded by paraffin wax so that the element in question was bombarded by slow neutrons. Under these conditions certain elements, notably lithium and boron, gave a very large increase in the number of counts which showed that a nuclear transformation was taking place.

Early in the following year the boron transformation was further investigated by Goldhaber in conjunction with H. J. Taylor. A photographic plate was soaked in a solution of borax and dried, thus introducing boron into the gelatine. The plate was then exposed for 15 hours to an 80 millicurie radon–beryllium source contained in a thick lead cylinder to reduce the effect of gamma radiation. Both source and plate were surrounded by paraffin. Under high magnification the plate revealed some 50 000 short straight tracks per square centimetre of its surface. Control experiments with untreated plates showed that each such track could only be ascribed to disintegration of a boron atom by a slow neutron with the emission of a single particle. Consideration of nuclear weights gave the reaction

$$^{10}_{5}B + ^{1}_{0}n \rightarrow ^{7}_{3}Li + ^{4}_{2}He$$

as the only one leading to emission of a single particle with release of energy. Similar experiments using a salt of lithium gave tracks indicating the emission of two particles, thus confirming the reaction already proposed by Chadwick and Goldhaber

$$^{6}_{3}Li + ^{1}_{0}n \rightarrow ^{4}_{2}He + ^{3}_{1}H$$

In the absence of particle emission, a nucleus capturing a slow neutron may be expected to emit a gamma ray of energy equal to the binding energy of the added neutron. This emission was demonstrated by Fermi and his colleagues with the aid of the apparatus depicted in Figure 13.1. A polonium–beryllium neutron source S of about 60

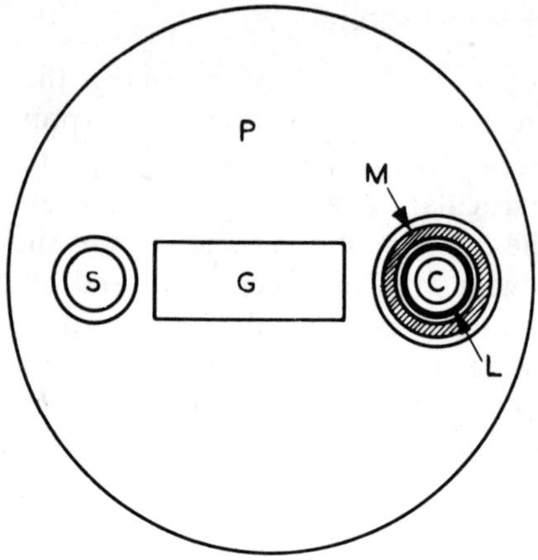

Figure 13.1 Demonstration of gamma rays from slow neutron capture

millicuries was immersed in a paraffin block P together with a Geiger–Müller counter C. A lead block G, 100 mm thick, was inserted into the paraffin between source and counter to shield the counter against direct gamma radiation from the source. The counter was surrounded by a 2 mm thickness of lead foil L and the count recorded. Small cylindrical shells of various materials M were then placed round the outside of the lead foil, and a marked increase in counts was generally observed when the materials were known to capture neutrons, notable exceptions being boron and lithium. No such increase in counts occurred with the paraffin removed, thereby confirming that the gamma rays giving rise to the increase were due to slow neutrons.

Although unable to project protons, slow neutrons could be detected by means of their reactions with boron and lithium, as suggested by Chadwick and Goldhaber. Early in 1935, Fermi and his co-workers described a detector employing this means in the form of an ionisation chamber filled with boron trifluoride. An alternative detector used by J. R. Dunning, G. B. Pegram, G. A. Fink and D. P. Mitchell consisted of an ionisation chamber fitted with a lithium front which was improved in 1935 by lining the chamber with lithium throughout. A chamber of this form gave excellent results in detecting the neutrons emitted by a radon–beryllium source at the centre of a 120 mm diameter sphere of paraffin. Practically all the neutrons, however, could be prevented from reaching the chamber by placing before it a layer of cadmium, 0.3 to 0.75 mm in thickness.

In order to determine whether the effect of cadmium on slow neutrons was mainly one of scattering or absorption, Dunning and his collaborators performed an experiment similar to that described in Section 12.4 for investigating the distribution of scattering angle. In this case, however, the source was situated at the centre of a 120 mm diameter sphere of paraffin and the ionisation chamber was lithium lined. Also the tapered platinum cylinder of the previous experiment was replaced by an assembly of stepped copper cylinders and cadmium discs. The materials tested were in the form of cylindrical shells instead of toroids but their axes were similarly aligned. The background count was first recorded without a shell of material in position. A copper shell between source and chamber increased the counts nearly threefold, but its replacement by a cadmium shell reduced the counts to slightly below the background level. It thus appeared that copper, which was known to obstruct a beam of slow neutrons far less than cadmium, was nevertheless able to scatter a fair proportion of the neutrons leaving the source towards the chamber. Cadmium, on the other hand, clearly did not scatter neutrons in this way and must even have prevented some of the neutrons scattered by the surroundings from reaching the chamber. The effect of cadmium on slow neutrons was thus seen to be almost entirely one of absorption and not of scattering.

13.3 Energy of slow neutrons

If neutrons are slowed down to an average energy equal to that of the surrounding atoms or molecules in thermal equilibrium, then, with each scattering collision, a small gain in their energy is as probable as an equally small loss. In this way they behave as gas molecules so that their velocities may be expected to have the statistical distribution derived in 1860 by James Clerk Maxwell from the kinetic theory of gases. According to this 'Maxwellian' distribution, the number of neutrons per unit volume having velocities between v and $v + \mathrm{d}v$ is

$$n'(v)\mathrm{d}v = 4\pi n(M_\mathrm{n}/2\pi kT)^{3/2}\, v^2 \exp(-M_\mathrm{n}v^2/2kT)\mathrm{d}v$$

where n is the total number of neutrons per unit volume, M_n the mass of the neutron, T the absolute temperature, and k Boltzmann's constant equal to 8.617×10^{-5} eV/°K. The dotted curve in Figure 13.2 gives $n'(v)/n$ as a function of v for a temperature of 20°C. Alternatively, if $n'(E)$ is the number of neutrons per unit volume having energies between E and $E + \mathrm{d}E$, the Maxwellian distribution becomes

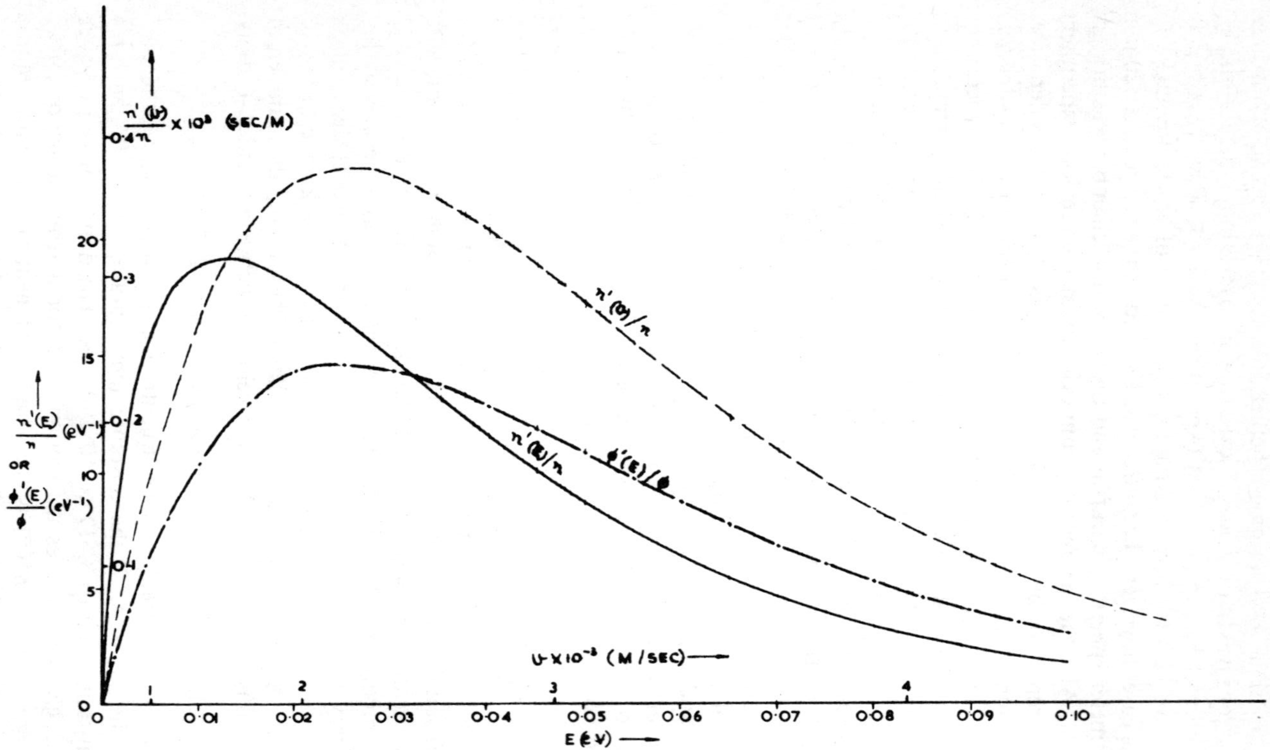

Figure 13.2 Maxwellian Distribution

$$n'(E)\mathrm{d}E = \frac{2\pi n}{(\pi kT)^{3/2}} E^{1/2} \exp(-E/kT)\mathrm{d}E \qquad (37)$$

and $n'(E)/n$ is given as a function of E by the full-line curve in Figure 13.2, also for a temperature of 20°C.

The belief that the terminal energy of slow neutrons followed the Maxwellian distribution received support in 1935 from Dunning and his co-workers, using the apparatus outlined in Figure 13.3. This apparatus was a form of 'mechanical velocity selector' and consisted of four discs made of duralumin, which readily transmits neutrons. Two of the discs AA, were spaced 540 mm apart and could be rotated by the driving shaft D. The other two discs, BB, were fixed, being spaced at 5 mm from the rotating discs. Fifty cadmium sectors extended round the whole of each disc, the circumferential width of the sectors being almost equal to that of the gaps between them. A radon–beryllium source S was located at the centre of a paraffin cylinder, 160 mm in diameter by 220 mm long. This cylinder was surrounded by a cadmium shield C, except for one face, which was placed close to one of the rotating discs. Close to the other rotating disc and directly opposite the source was a lithium-lined ionisation chamber I.

The operation of a mechanical velocity selector of this type has been lucidly explained by G. A. Fink. With the cadmium sectors on one of the fixed discs immediately opposite those on the other, and likewise the sectors on the rotating discs, the area through which neutrons can pass beyond the fixed disc nearer the source varies with rotating sector travel as shown by the full line in Figure 13.4. By the time a neutron of

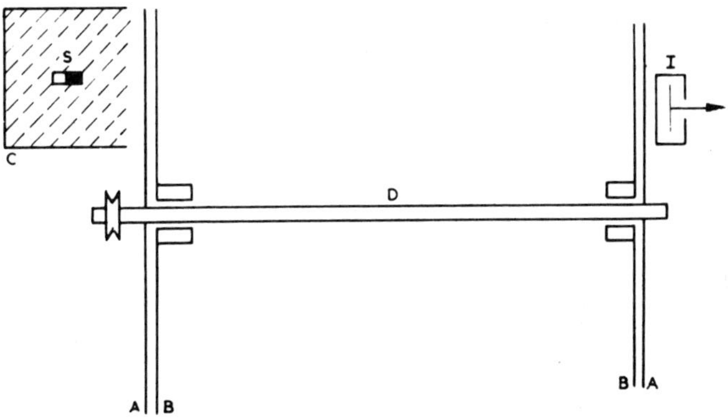

Figure 13.3 Mechanical velocity selector

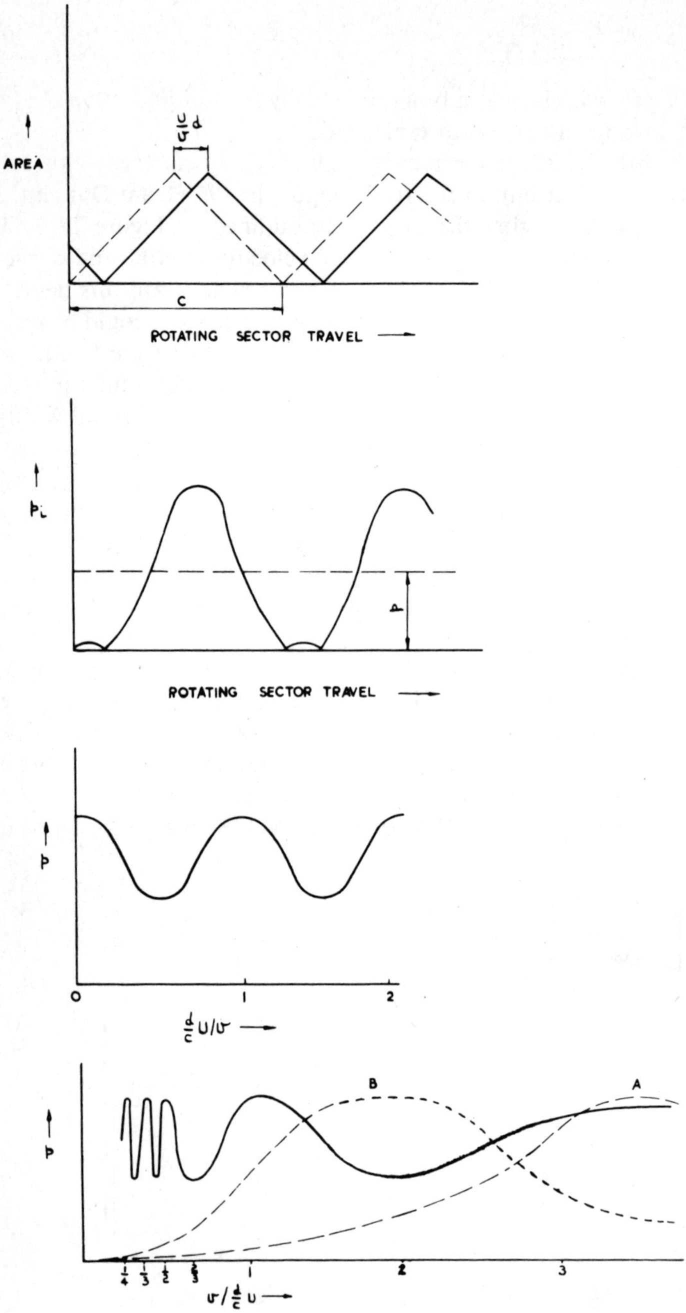

Figure 13.4 Neutron transmission curves for mechanical velocity selector

velocity v has covered the distance d between the fixed discs, the rotating sectors will have travelled a distance Ud/v where U is the sector velocity. In terms of the sector travel at the rotating disc nearer the source, therefore, the area through which neutrons can pass beyond the other rotating disc will be as shown by the dotted line in Figure 13.4. Furthermore, for a given travel of the sectors nearer the source, the product of the areas through which a neutron can pass at either end of the selector is proportional to the corresponding probability p_i of neutrons reaching the ionisation chamber. The probability p_i varies with sector travel as shown in Figure 13.4, and its mean value p is directly proportional to the fraction of incident neutrons transmitted by the selector.

The fraction p attains its maximum value whenever Ud/v is an integral multiple of the circumferential pitch c of the sectors and falls to its minimum value when Ud/v lies half-way between successive such multiples. The variation of p with Ud/vc will therefore be as shown in Figure 13.4, from which the corresponding variation of p with vc/Ud can be derived. At slow rotational speeds, the neutron velocity v will correspond to a high value of vc/Ud and the neutron velocity distribution will be given by a curve such as the dotted one A. If the ordinate of this distribution curve is then everywhere multiplied by the corresponding value of p, the area under the resulting curve will be proportional to the transmitted fraction of incident neutrons of all velocities. With increasing rotational speed, therefore, the peak of the distribution curve will move to lower values of vc/Ud and the transmitted fraction will fall, reaching a minimum when the peak of the distribution curve has attained a position such as B, where vc/Ud is approximately equal to 2.

As the rotational speed of the selector was increased, the counts were in fact found to fall to a minimum, reaching it at a speed of about 2500 r.p.m. The neutron distribution curve would have thus been expected to attain its peak at a neutron velocity of

$$\begin{aligned} 2 \times 540 \times 50 \times 2500/60 &= 2\,250\,000 \text{ mm/sec} \\ &= 2250 \text{ m/sec} \end{aligned}$$

As Figure 13.2 shows, this velocity agrees closely with that at which the peak occurs in a Maxwellian distribution, and the assumption of this distribution for fully slowed down neutrons was thereby confirmed. It followed that the average energy of such neutrons was indeed that of the surrounding atoms in thermal equilibrium and, as a result, a neutron that has been slowed down to this energy is referred to as a 'thermal' neutron and the corresponding slowing-down process as 'thermalisation'.

On differentiating the Maxwellian distribution $n'(v)$ with respect to v and equating the result to zero, the distribution curve is found to reach its peak at a velocity $(2kT/M_n)^{1/2}$, known as the 'most probable' velocity. At room temperature this velocity is 2200 m/sec. The energy at the most probable velocity is merely kT, and at 20 °C, the room temperature commonly adopted, this energy becomes 0.0252 eV, although for many purposes it is taken simply as 0.025 eV. The average energy over the Maxwellian spectrum is 50% greater than this but it is customary to quote the value 0.025 eV when referring to the energy of thermal neutrons.

13.4 Total cross-sections for thermal neutrons

An early determination of total cross-sections for thermal neutrons was made by Dunning and his colleagues in 1935, using the apparatus outlined in Figure 13.5. A radon–beryllium neutron source S was located at the centre of a 120 mm diameter paraffin sphere P. The effect of room scattering was minimised by means of a cylindrical cadmium box B round the paraffin sphere and a cadmium shield C round the lithium-lined ionisation chamber I. The materials M investigated covered 73 elements. They took the form of cylindrical samples, having a diameter of 40 mm and a thickness such that the proportion of incident thermal neutrons transmitted was 40 per cent. A cadmium diaphragm H defined the neutron beam to the sample size, and a 3 mm thick lead disc G formed a gamma ray shield. Thermal neutron transmission to the chamber was measured by first taking a count with a cadmium disc D in position, and then subtracting this count from a second one taken with the disc removed. As in the determination of fast neutron cross-sections, described in Section 12.1, the total cross-section for thermal neutrons was then obtained from the relationship

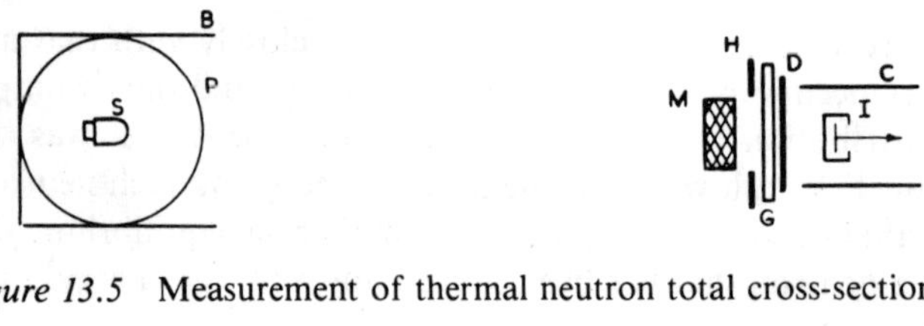

Figure 13.5 Measurement of thermal neutron total cross-section

$$\Sigma_t = \frac{1}{T} \log \left(\frac{\text{thermal neutron transmission to chamber without sample}}{\text{thermal neutron transmission to chamber with sample}} \right)$$

where T was the sample thickness.

With few exceptions, the total cross-section for thermal neutrons was found to exceed that of the same material for fast neutrons, the excess being remarkably great in the case of a few elements such as cadmium. Furthermore, as described in Section 13.2, the effect of cadmium on thermal neutrons had been found to be mainly one of absorption. It thus appeared to be by its absorption component that the total cross-section for thermal neutrons could greatly exceed that of the same material for fast neutrons, its scattering component and the total cross-section for fast neutrons being supposedly of similar magnitude. Although this conclusion has since been generally confirmed, the experiments on which it was based were crude by modern standards. They gave cross-sections half of which were some 10 to 20 per cent lower than present-day accepted values, the remainder differing even more widely.

In more refined experiments for determining neutron total cross-sections, the proportion of scattered neutrons entering the detector is reduced to a minimum not only by ensuring that the distances between source, detector and material investigated are sufficiently large but also by the introduction of small aperture collimating diaphragms on either side of the material.

Slow neutrons of selected energies can be obtained by means of improved forms of the velocity selector described in Section 13.3. In these selectors, the rate at which neutrons can leave the source for the detector does not vary with time in the saw-tooth manner shown in Figure 13.4 but reaches its maximum almost instantaneously and, after a very short time interval, decreases as rapidly to zero, remaining at this level for a comparatively long period of time t_p until the cycle is repeated. The sensitivity of the detector varies with time in a similar manner but reaches its peak later by a time interval τ. The probability p_i of neutrons reaching the detector and activating it is therefore zero unless their velocity v is very close to one of the values,

$$d/\tau,\ d/(t_p + \tau),\ d/(2t_p + \tau),\ d/(3t_p + \tau),\ \ldots$$

where d is the distance from source to detector. If the period of time t_p is made sufficiently large, all except the first of these velocities will be very low and thus apply to a relatively small number of the neutrons activating the detector. The bulk of the neutrons counted will then be of nearly uniform energy, the magnitude of which can be varied by adjusting the time interval τ. If at each such interval counts are taken with and without a sample of material between source and detector, the total cross-section of the material can be obtained as a function of energy from about 0.01 to 1000 eV.

13.5 Scattering of thermal neutrons

Over the years 1935 and 1936 information was obtained on scattering cross-sections for slow neutrons, notably by A. C. G. Mitchell and E. J. Murphy. In these investigations a thin metal detector was activated by placing it against the face of a paraffin block from which slow neutrons were emitted. The detector was covered with a thin layer of the scattering material to be examined so that some of the neutrons passing through the detector were reflected back into it. The consequent increase in activation on introducing the scattering material then gave a measure of its scattering cross-section. The results obtained in this way were subject to serious uncertainties and existed for comparatively few materials when in 1937, M. Goldhaber and G. H. Briggs reported experiments on scattering cross-sections for thermal neutrons that not only extended over a wide range of elements but employed a technique closely allied to that used today.

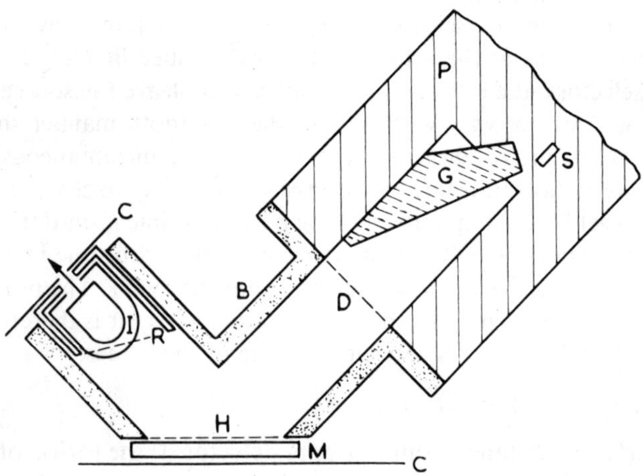

Figure 13.6 Measurement of thermal neutron scattering cross-section

The experimental arrangement used by Goldhaber and Briggs is shown diagrammatically in Figure 13.6. A 60 millicurie radium–beryllium neutron source S was embedded in paraffin P, the neutron beam being defined by mutually perpendicular cylindrical shells of boron oxide B. A sheet of the scattering material M was placed to cover an elliptical opening H in the boron oxide so that neutrons struck the material at an angle of about 45°. Such of these neutrons as were scattered through an angle of 90° or so remained within the boron oxide to enter an ionisation chamber I filled with boron trifluoride at 2 atmospheres pressure. To increase the number of neutrons recorded, the back and sides of the chamber were covered with a paraffin wax reflector R about 2.5 mm thick. The chamber was protected from the effects of room scattering both by the boron oxide shells and by sheets of cadmium C, whilst a lead cone G shielded it against direct gamma radiation from the source.

With a 1 mm thick cadmium disc at D, the chamber count rate included no contribution from thermal neutrons leaving the paraffin. This count rate was subtracted from that without the cadmium disc to give a rate I due only to thermal neutrons leaving the paraffin. The scattered neutrons from all scattering elements other than hydrogen were assumed to have the same angular distribution. As pointed out in Section 11.4, the probability of a neutron making a collision in travelling a small distance ds is $\Sigma \mathrm{d}s$, where Σ is the appropriate cross-section. Denoting, therefore, any two scattering elements by suffixes A and B, their scattering cross-sections by Σ_s and their thickness by T,

$$\Sigma_{sA}/\Sigma_{sB} = (\mathrm{d}I_A/\mathrm{d}T_A)_{T_A=0}/(\mathrm{d}I_B/\mathrm{d}T_B)_{T_B=0}$$

For comparison with the scattering cross-sections, total cross-sections were obtained by means of an experimental technique similar to that used by Dunning and his collaborators but improved to give more reliable results. In fact, of the 11 total cross-sections so found, 9 were within 10 per cent of present-day values. It was now seen that the ratio of the total cross-section for paraffin to that for carbon was within a few per cent of the corresponding ratio of their scattering cross-sections, as determined from the above equation for Σ_{sA}/Σ_{sB}. It was thus concluded that the total and scattering cross-sections were the same for either paraffin or carbon so that the scattering cross-section Σ_{sA} for any element A could be obtained by taking element B as carbon, with Σ_{sB} equal to its known total cross-section. Of the 61 elements investigated, 47 gave sufficiently high count rates for values of scattering cross-section to be assigned to them and of these 33 were within 20 per cent of the values accepted today.

Scattering cross-sections cannot be determined as accurately as can total cross-sections. Since the scattered neutrons that enter the detector must travel for longer distances within the material investigated than unscattered neutrons, then, in order to avoid multiple scattering, the thickness of the material must be smaller than that suited to the measurement of total cross-sections, with the result that most of the incident neutrons are transmitted without collision. Furthermore, of the few neutrons that undergo scattering, only a small proportion do so through an angle that enables them to enter the detector. A strong neutron source is therefore essential for measurements of this kind. In principle, the absorption cross-section can be determined as the difference between the total and scattering cross-sections, but uncertainty in the scattering cross-section is generally such that this determination is only of value when the total and scattering cross-sections differ considerably.

13.6 The 1/*v* Absorption Law

A theoretical explanation of the very large thermal neutron cross-sections observed for certain elements was given by H. A. Bethe in 1935. From the supporting analysis it appeared that absorption cross-sections for slow neutrons should vary inversely as the velocity of neutron relative to absorbing nucleus. Later in the same year, this theory was pursued experimentally by F. Rasetti, E. Segré, G, Fink, J. R. Dunning and G. B. Pegram. A 500 mm diameter steel disc was coated with a cadmium film 0.02 mm thick and rotated at 6000 r.p.m. A beam of thermal neutrons was directed at an angle of about 25° to the face of the disc near its edge, intersecting a radius of the disc at right angles. When the disc was rotated so that the neutron velocity component in the plane of the disc had the same direction as the local velocity of the disc itself, then both the velocity of the neutrons relative to the disc and the length of their relative path through it were reduced in the same proportion. With the disc rotated in the opposite direction, the relative velocity and path length were proportionally increased. The neutrons passing through the disc were recorded by a lithium-lined ionisation chamber, and a reduction in count rate was noted when the rotation of the disc was reversed so as to increase the relative neutron velocity. With the disc stationary, the number of absorption mean free paths through it was known so that the reduction in count rate on reversing the rotation could be expressed as a percentage increase in number of mean free paths. This increase was seen to be only slightly less than the corresponding increase in the relative path length, as found by taking the neutron absolute velocity equal to the average over a Maxwellian spectrum at room temperature. The conclusion followed that the absorption cross-section of cadmium for thermal neutrons was practically independent of relative neutron velocity.

Although cadmium did not show the expected inverse variation of slow neutron absorption cross-section with relative neutron velocity, other elements were found to do so. When the disc was replaced by one of silver 0.8 mm thick, the change in count rate on reversing the rotation was less than the experimental error. Further counts were taken with a disc of aluminium, which was coated with boron carbide of a thickness sufficient to absorb some 60 per cent of the neutrons incident upon it. In this case the change in count rate on reversing the rotation was only slightly greater than the experimental error. The substantially unchanged count rate was consistent with there being no change in the number of mean free paths with increase in relative neutron velocity and path length. This result pointed to an absorption cross-section for thermal neutrons in silver and boron that varied inversely as the relative neutron velocity, a variation known as the '$1/v$ law'.

For nearly all materials, absorption cross-sections for thermal neutrons have been found to follow the $1/v$ law, and are conventionally quoted at a neutron energy corresponding to the most probable velocity in a Maxwellian distribution (0.025 eV at 20°C). As pointed out by H. H. Goldsmith and F. Rasetti in 1936, this is the energy of a parallel beam of monoenergetic neutrons that undergoes 50 per cent transmission through the same thickness of $1/v$ absorber as a beam of neutrons conforming to the Maxwellian distribution. On the other hand, scattering cross-sections often vary in an irregular way at thermal energies so that average values over the Maxwellian distribution are of more interest than values at a particular energy. Scattering and absorption cross-sections for thermal neutrons are given in Appendix 5.

14

Diffusion of neutrons

14.1 Flux

The passage of neutrons through a material has so far been studied only for cases in which the neutrons enter as a parallel beam. Moreover, only such of those neutrons as have thereafter avoided collision up to the point considered in the material have been taken into account. A more general treatment must clearly allow for neutrons entering the material from all directions and must take account of neutrons either leaving the material or being absorbed within it after making several scattering collisions. With evidence that the energy distribution of thermal neutrons followed Maxwell's prediction from the kinetic theory of gases, it was natural that a theoretical treatment of the passage of thermal neutrons through matter should follow the lines of that developed for gas molecules. Such a treatment was attempted by E. Fermi in 1936 and, together with E. Amaldi, he was able to report results that accorded well with experiment. Since that time the theory has been further refined and its place in the study of neutrons is now firmly established.

In its present form as applied to neutrons, the theory employs the concept of 'flux'. This concept may be described with the aid of Figure 14.1, in which any direction is represented by a straight line in that direction through the centre of a sphere of unit radius. Directions within a small range about that one are then defined by a small area of the surface of the sphere, or increment of solid angle $d\Omega$ in which the representative straight lines are contained. The surface area of the sphere is imagined to be made up of such increments, each of which is added in a prescribed order to leave the area Ω so far formed continuous. The increment $d\Omega$ is the last to be added so that the area, or solid angle, Ω defines the direction about which $d\Omega$ ranges.

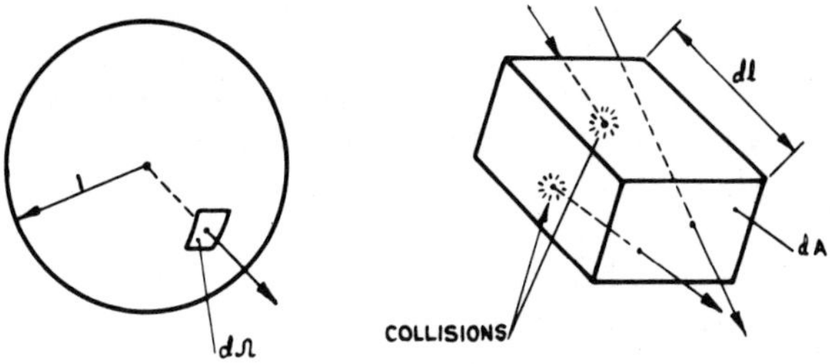

Figure 14.1 Neutron flux

Consideration will first be limited to neutrons that travel within the small range of solid angle $d\Omega$ about direction Ω and within the small range dv about velocity v, corresponding to the small range dE about energy E. At any point in the medium through which such neutrons are travelling, their density will be written $n^{*\prime}(\Omega, E)\, d\Omega\, dE$. A volume element surrounding this point will be taken to have a cross-sectional area dA normal to the direction Ω and a length in that direction equal to dl. Of the neutrons within the ranges $d\Omega$ and dE, some will enter the element and terminate their existence before leaving it as the result of collisions in which they are either absorbed, or scattered outside the ranges $d\Omega$ or dE. Others will originate within the element as the result of collisions that scatter them into the ranges $d\Omega$ and dE. The remainder will pass through the element unchanged in direction or energy. If the density $n^{*\prime}\, d\Omega\, dE$ is not to vary with time, then for every neutron that ceases to contribute to it within the element another one must begin to do so.

If $(\Sigma)_E$ is the cross-section of the medium for a particular form of collision then in a short interval of time dt each neutron makes a number of such collisions equal to $(\Sigma)_E v\, dt$. The rate at which neutrons within the ranges $d\Omega$ and dE make such collisions in the volume element is therefore

$$(\Sigma)_E v n^{*\prime}(\Omega, E) d\Omega\, dE\, dA\, dl$$

so that

$$\text{rate of collision/unit volume} = (\Sigma)_E v n^{*\prime}(\Omega, E) d\Omega\, dE \qquad (38)$$

The quantity $vn^{*\prime}(\Omega, E)$ appearing in this equation may be considered in the following way. In a short interval of time dt, the neutrons within unit solid angle about direction Ω and within unit range about energy E that leave the volume element amount to

$$v\,\mathrm{d}t\,.\,\mathrm{d}A\ n^{*\prime}(\Omega, E)$$

Since these neutrons leave the element by the face of area dA, the rate at which they cross unit area normal to direction Ω is

$$vn^{*\prime}(\Omega, E)$$

which is the quantity being considered. This quantity is, then, the rate at which neutrons, falling within unit range about energy E and travelling within unit solid angle about direction Ω, cross unit area normal to that direction. It is called the 'angular flux per unit energy' and denoted by the symbol $\phi^{*\prime}$. Equation (38) may thus be written in either of the alternative forms

$$\left.\begin{aligned}\text{rate of collision/unit volume} &= (\Sigma)_E v n^{*\prime}(\Omega, E)\mathrm{d}\Omega\,\mathrm{d}E\\ &= (\Sigma)_E \phi^{*\prime}(\Omega, E)\mathrm{d}\Omega\,\mathrm{d}E\end{aligned}\right\} \quad (39)$$

for all neutrons within range dE about energy E and solid angle dΩ about direction Ω.

For neutrons within the same energy range but travelling in all directions, equations (39) give

$$\left.\begin{aligned}\text{rate of collision/unit volume} &= (\Sigma)_E v\left[\int_0^{4\pi} n^{*\prime}(\Omega, E)\mathrm{d}\Omega\right]\mathrm{d}E\\ &= (\Sigma)_E\left[\int_0^{4\pi} \phi^{*\prime}(\Omega, E)\mathrm{d}\Omega\right]\mathrm{d}E\end{aligned}\right\} \quad (40)$$

The quantity $\int_0^{4\pi} n^{*\prime}(\Omega, E)\mathrm{d}\Omega$ is the density of neutrons falling within unit range about energy E and travelling in all possible directions. It is identical with the quantity $n'(E)$ appearing in Section 13.3. Likewise, the quantity $\int_0^{4\pi} \phi^{*\prime}(\Omega, E)\mathrm{d}\Omega$ is the rate at which neutrons, falling within unit range about energy E and travelling within a small range of solid angle about a given direction, cross unit area normal to that direction, summed over all possible directions. This quantity is called the 'flux per unit energy' and is denoted by the symbol ϕ'. Equations (40) may thus be written

$$\left.\begin{aligned}\text{rate of collision/unit volume} &= (\Sigma)_E v n'(E)\mathrm{d}E\\ &= (\Sigma)_E \phi'(E)\mathrm{d}E\end{aligned}\right\} \quad (41)$$

for all neutrons within range dE about energy E.

Finally, for all neutrons of a group having a particular energy distribution or 'spectrum', equations (41) give

rate of collision/unit volume

$$\left.\begin{aligned} &= \int_0^\infty (\Sigma)_E v n'(E)\mathrm{d}E = \Sigma\bar{v}\int_0^\infty n'(E)\mathrm{d}E \\ &= \int_0^\infty (\Sigma)_E \phi'(E)\mathrm{d}E = \Sigma\int_0^\infty \phi'(E)\mathrm{d}E \end{aligned}\right\} \quad (42)$$

where $\bar{v}$, Σ are respectively the average velocity and cross-section over the spectrum. The quantity $\int_0^\infty n'(E)\mathrm{d}E$ is the density of all neutrons falling within a particular energy range and travelling in all possible directions. It is identical with the quantity n appearing in Section 13.3. Likewise, the quantity $\int_0^\infty \phi'(E)\mathrm{d}E$ is the rate at which neutrons within a particular energy range, travelling within a small range of solid angle about a given direction, cross unit area normal to that direction, summed over all possible directions. This quantity is called the 'flux' and is denoted by the symbol ϕ. Finally, therefore, equations (42) may be written

$$\text{rate of collision/unit volume} \left.\begin{aligned} &= \Sigma\bar{v}n \\ &= \Sigma\phi \end{aligned}\right\} \quad (43)$$

for all neutrons of the group.

In the case of an element having a cross-section varying inversely as neutron velocity, the cross-section at neutron energy E is

$$(\sigma)_E = (\sigma)_{E^*}\sqrt{(E^*/E)}$$

where $(\sigma)_{E^*}$ is the tabulated cross-section at energy E^*. The second of equations (42) then gives

$$\begin{aligned} \sigma &= (\sigma)_{E^*}\sqrt{(E^*)}\int_0^\infty \frac{\phi'(E)}{\sqrt{E}}\,\mathrm{d}E \bigg/ \int_0^\infty \phi'(E)\mathrm{d}E \\ &= (\sigma)_{E^*}\sqrt{(E^*)}\int_0^\infty \frac{\phi'(E)/\phi}{\sqrt{E}}\,\mathrm{d}E \end{aligned} \quad (44)$$

If the neutrons belong to the thermal group, conforming to the Maxwellian spectrum, then $n'(E)\mathrm{d}E$ is given by equation (37) so that

$$\phi'(E)/\phi = vn'(E) \Big/ \int_0^\infty vn'(E)\mathrm{d}E$$

$$= [\sqrt{E}\, n'(E)/n] \Big/ \int_0^\infty \sqrt{E}\, \frac{n'(E)}{n}\, \mathrm{d}E$$

$$= \sqrt{E}\, \frac{n'(E)}{n} \Big/ \left[\frac{2\pi}{(\pi kT)^{3/2}} \int_0^\infty E\mathrm{e}^{-E/kT}\mathrm{d}E\right]$$

On integrating by parts,

$$\phi'(E)/\phi = \tfrac{1}{2}\sqrt{(\pi/kT)}\, \sqrt{E}\, n'(E)/n$$

$$= (1/kT)^2\, E\mathrm{e}^{-E/kT} \qquad (45)$$

and is shown plotted against E by the chain-dotted curve of Figure 13.2. It will be noted $\phi'(E)/\phi$ peaks at the energy corresponding to the most probable velocity. Introducing equation (45) into (44), the average cross-section for the thermal group becomes

$$\sigma = (\sigma)_{E^*}\, \frac{\sqrt{(E^*)}}{(kT)^2} \int_0^\infty E^{\frac{1}{2}}\, \mathrm{e}^{-E/kT}\, \mathrm{d}E$$

Substituting x^2 for E/kT, the integral may be written

$$2(kT)^{3/2} \int_0^\infty x^2\, \mathrm{e}^{-x^2}\, \mathrm{d}x = -(kT)^{3/2} \left[x\, \mathrm{e}^{-x^2} - \int \mathrm{e}^{-x^2}\, \mathrm{d}x\right]_0^\infty$$

On introducing the limits, the first term within the square brackets vanishes whilst the second becomes the normal error integral, having the value $\frac{1}{2}\sqrt{\pi}$. Finally, therefore,

$$\sigma = \tfrac{1}{2}\sqrt{\pi}\, (\sigma)_{E^*}\, \sqrt{(E^*)}/\sqrt{(kT)}$$

The tabulated cross-section $(\sigma)_{E^*}$ will generally refer to the energy at the most probable velocity, in which case

$$E^* = kT$$

and the average cross-section becomes

$$\sigma = \tfrac{1}{2}\sqrt{\pi}\, (\sigma)_{E^*} \qquad (46)$$

14.2 Current density

In order to find the variation in flux of thermal neutrons as they pass through a medium, the rate at which they are generated in a unit volume of the medium and scattered into it is set equal to the rate at which they are absorbed in it and scattered out of it. The rate at which neutrons are generated in a unit volume will be denoted by S and will frequently be zero. The rate at which the neutrons are absorbed in a unit volume is seen from equation (43) to be merely

$$\Sigma_a \phi \tag{47}$$

where Σ_a is the absorption cross-section. The rate at which neutrons are scattered into or out of a unit volume is more difficult to represent. Replacing Σ_a in expression (47) by the scattering cross-section Σ_s would give only the rate at which neutrons made scattering collisions within the unit volume and provide no representation of the rate at which they were scattered into or out of it.

The representation required, however, may be made in terms of the concept of 'current density' and the problem then reduces to one of expressing this concept in terms of flux. The current density in a given direction is the rate at which neutrons, having a component of velocity in that direction, pass from one side to the other of unit area normal to that direction. Current density is denoted by the symbol J, the direction to which it applies being indicated by a suffix. In Figure 14.2, for instance, an illustration is given of the current density J_{+x} in the positive direction of the coordinate x and of J_{-x} in the negative direction.

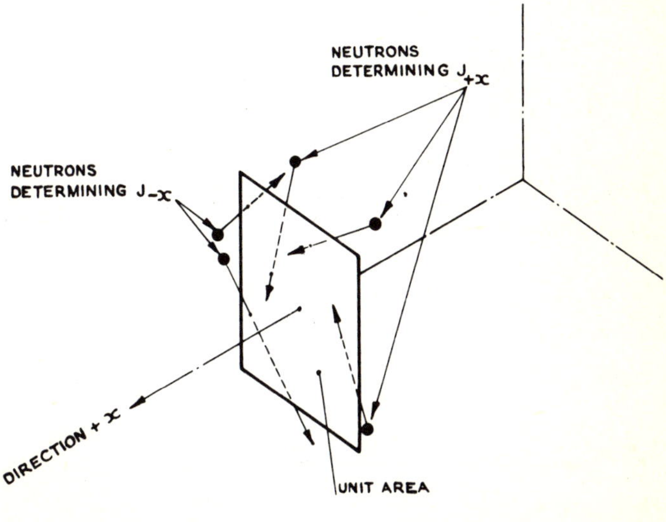

Figure 14.2 Current density

The 'nett current density' in a given direction is the current density in that direction less the current density in the opposite direction. It is denoted by J, conventionally taken to apply to the positive direction of the coordinate indicated by a suffix. In the case of the coordinate x shown in Figure 14.2, for instance,

$$J_x = J_{+x} - J_{-x}$$

The concept of current density may now be employed to express the rate at which neutrons are scattered into or out of a unit volume. Referring to Figure 14.3, the nett rate at which neutrons are scattered into the element of volume $dx\, dy\, dz$ across the three faces having a common corner at the point (x, y, z) is

$$J_x\, dy\, dz + J_y\, dz\, dx + J_z\, dx\, dy$$

Similarly, the nett rate at which neutrons are scattered out of the element across the other three faces is

$$\left(J_x + \frac{\partial J_x}{\partial x}\, dx\right) dy\, dz + \left(J_y + \frac{\partial J_y}{\partial y}\, dy\right) dz\, dx + \left(J_z + \frac{\partial J_z}{\partial z}\, dz\right) dx\, dy$$

On subtracting the first of these two rates from the second, the nett rate of neutron leakage from the element is found to be

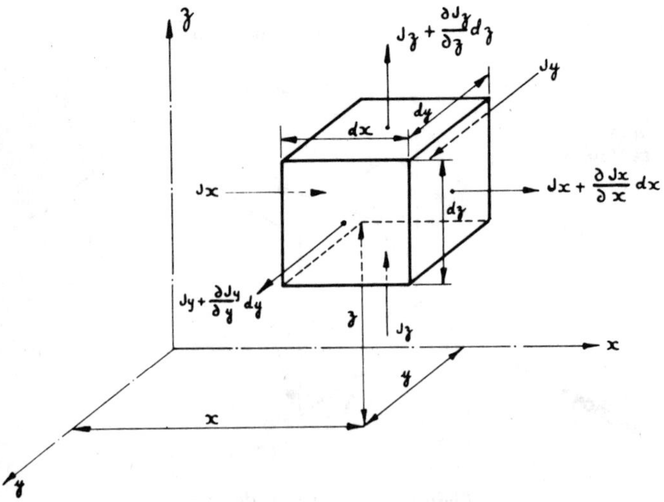

Figure 14.3 Neutron leakage

$$\left(\frac{\partial J_x}{\partial x} + \frac{\partial J_y}{\partial y} + \frac{\partial J_z}{\partial z}\right) \mathrm{d}x\, \mathrm{d}y\, \mathrm{d}z,$$

so that the nett rate of leakage from unit volume is

$$\frac{\partial J_x}{\partial x} + \frac{\partial J_y}{\partial y} + \frac{\partial J_z}{\partial z} \tag{48}$$

If ϕ and J are not to vary with time, then from results (47) and (48), the rate per unit volume at which neutrons are generated can be expressed, in the case of the thermal group, as

$$S = \Sigma_a \phi + \frac{\partial J_x}{\partial x} + \frac{\partial J_y}{\partial y} + \frac{\partial J_z}{\partial z} \tag{49}$$

14.3 Integral equation between flux and current density

In order to reduce equation (49) to one in terms of flux alone, the current density must now be related to the flux. For the current density in any particular direction, the required relation is obtained on expressing the contribution to that current density made by neutrons scattered from an element of volume in terms of the flux at that element and equating to the current density the sum of all such contributions throughout the scattering medium.

Referring to Figure 14.4, a small area $\mathrm{d}A$ lies normal to the z-axis at the point having Cartesian coordinates x, y, z. An element of volume $\mathrm{d}V_1$ is situated at the point (x_1, y_1, z_1) having spherical coordinates r, θ, φ referred to the point (x, y, z) as origin. The rate at which neutrons are scattered within the element $\mathrm{d}V_1$ is

$$\phi_1 \Sigma_s \mathrm{d}V_1 \tag{50}$$

where ϕ_1 is the local neutron flux. If the scattering is isotropic in the laboratory system so that a neutron leaves a collision centre in any direction with equal probability, then the probability that a neutron will be scattered from the volume element $\mathrm{d}V_1$ in the direction of the area $\mathrm{d}A$ is

$$\mathrm{d}A \cos\theta / 4\pi r^2 \tag{51}$$

Also, the probability that this neutron will reach the area $\mathrm{d}A$ without further collision is

$$\exp(-\Sigma_t r) \tag{52}$$

The rate at which neutrons are scattered from volume $\mathrm{d}V_1$ so as to reach area $\mathrm{d}A$ without further collision is therefore

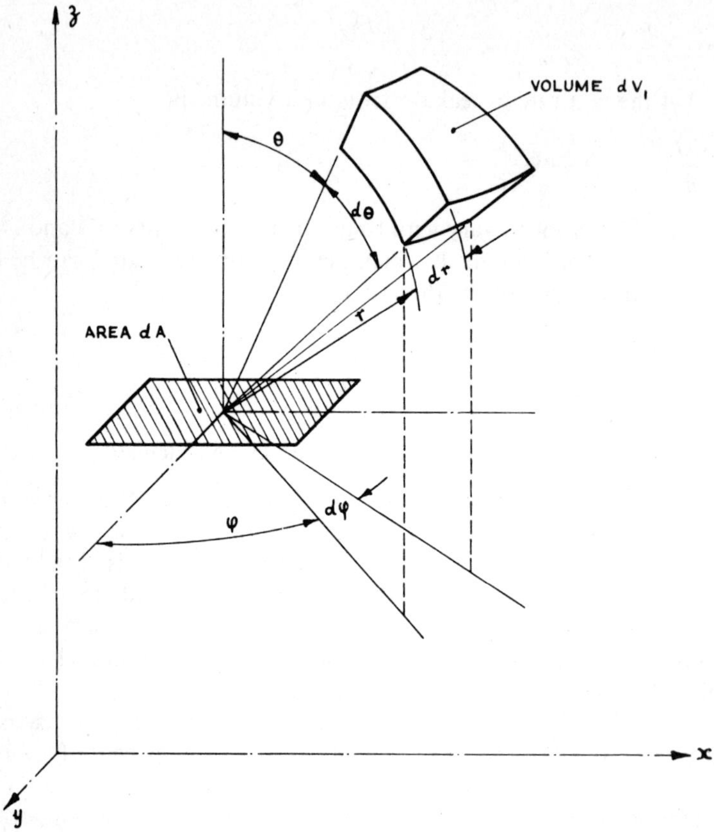

Figure 14.4 Diagram for relating flux to current density

$$\phi_1 \Sigma_s \, dV_1 \, dA(\cos\theta/4\pi r^2) \exp(-\Sigma_t r) \tag{53}$$

Any neutron that reaches area dA from volume dV_1 either after further collision or without collision in volume dV_1 is automatically accounted for when expression (53) is integrated over all volume elements for which $z_1 - z$ is positive. On carrying out this integration the result may be equated to J_{-z} dA. Introducing the relationship

$$dV_1 = r^2 \sin\theta \, d\varphi \, d\theta \, dr \tag{54}$$

the equation thus obtained is

$$J_{-z} = \tfrac{1}{4}(\Sigma_s/\Sigma_t) \int_0^{\pi/2} \int_{-\pi}^{\pi} \int_0^{\infty} \frac{\Sigma_t}{\pi} \phi_1 \exp(-\Sigma_t r) \cos\theta \sin\theta \, dr \, d\varphi \, d\theta \tag{55}$$

Since the integration has been carried out over the range of r from zero to infinity, equation (55) is strictly valid only for the case of an infinite medium. However, the term $\exp(-\Sigma_t r)$ becomes very small when r is greater than about three total mean free paths. As a result, the equation still holds fairly well up to such distances from the nearest boundary of a finite medium.

An integral equation similar to (55) was set up by R. Peierls in 1939. In Peierls's equation, however, the $\cos\theta$ term in the right-hand side was omitted and the upper limit in the first integral set equal to π. The resulting expression thus represented the rate at which neutrons from all volume elements dV_1 crossed unit area at the point (x, y, z) normal to their directions of travel. It could thus be equated to $\phi(x, y, z)$ to give an equation for the flux without the aid of the current density concept. Even so, the form of the integral equation given here is more convenient when expressions for current densities as well as flux are nevertheless still required.

In order to carry out the integrations in equation (55), ϕ_1 may be expressed in the form of a Taylor series and will then be the sum of all terms such as

$$\frac{1}{a!b!c!}\left(\frac{\partial^{a+b+c}\phi}{\partial x^a\,\partial y^b\,\partial z^c}\right)(x_1 - x)^a(y_1 - y)^b(z_1 - z)^c \tag{56}$$

where a, b, c may each have any of the values 0, 1, 2, 3, etc. Operators such as $\partial^0/\partial x^0$ are regarded as multiplying by unity and the first term for instance, in which a, b and c are each zero, then reduces as it should to ϕ.

On substituting expression (56) for ϕ_1 in equation (55) and introducing the relations

$$\left.\begin{aligned} x_1 - x &= r\sin\theta\sin\varphi \\ y_1 - y &= r\sin\theta\cos\varphi \\ z_1 - z &= r\cos\theta \end{aligned}\right\} \tag{57}$$

the triple integral in (55) becomes the sum of expressions such as

$$\frac{1/\pi}{a!b!c!}\left(\frac{\partial^{a+b+c}\phi}{\partial x^a\,\partial y^b\,\partial z^c}\right)\int_0^{\pi/2}\int_{-\pi}^{\pi}\int_0^{\infty}\Sigma_t r^{a+b+c}\exp(-\Sigma_t r)$$

$$\sin^a\varphi\cos^b\varphi\sin^{a+b+1}\theta\cos^{c+1}\theta\,dr\,d\varphi\,d\theta \tag{58}$$

The terms containing r may be integrated with respect to r to give

$$\int_0^{\infty}\Sigma_t r^{a+b+c}\exp(-\Sigma_t r)\,dr = -\left[r^{a+b+c}\exp(-\Sigma_t r)\right]_0^{\infty} +$$

$$(a+b+c)\int_0^\infty r^{a+b+c-1}\exp(-\Sigma_t r)\mathrm{d}r$$

The first term on the right-hand side is zero so that successive integration yields the result

$$\int_0^\infty \Sigma_t r^{a+b+c}\exp(-\Sigma_t r)\,\mathrm{d}r = (a+b+c)!/\Sigma_t^{a+b+c} \tag{59}$$

Integrating likewise the terms containing φ with respect to φ,

$$\int_{-\pi}^{\pi}\sin^a\varphi\cos^b\varphi\,\mathrm{d}\varphi = \left[\frac{\sin^{a+1}\varphi\cos^{b-1}\varphi}{a+b}\right]_{-\pi}^{\pi} + \frac{b-1}{a+b}\int_{-\pi}^{\pi}\sin^a\varphi\cos^{b-2}\varphi\,\mathrm{d}\varphi$$

The first term on the right-hand side is again zero so that successive integration gives

$$\left.\begin{aligned}\int_{-\pi}^{\pi}\sin^a\varphi\cos^b\varphi\,\mathrm{d}\varphi &= \left(\frac{b-1}{a+b}\right)\left(\frac{b-3}{a+b-2}\right)\cdots\left(\frac{1}{a+2}\right)\times\\ &\quad\int_{-\pi}^{\pi}\sin^a\varphi\,\mathrm{d}\varphi,\ \text{if } b \text{ is even}\\ &= \left(\frac{b-1}{a+b}\right)\left(\frac{b-3}{a+b-2}\right)\cdots\left(\frac{2}{a+3}\right)\times\\ &\quad\int_{-\pi}^{\pi}\sin^a\varphi\cos\varphi\,\mathrm{d}\varphi,\ \text{if } b \text{ is odd}\end{aligned}\right\} \tag{60}$$

Now

$$\int_{-\pi}^{\pi}\sin^a\varphi\,\mathrm{d}\varphi = -\left[\frac{\sin^{a-1}\varphi\cos\varphi}{a}\right]_{-\pi}^{\pi} + \frac{a-1}{a}\int_{-\pi}^{\pi}\sin^{a-2}\varphi\,\mathrm{d}\varphi$$

$$\left.\begin{aligned}&= \left(\frac{a-1}{a}\right)\left(\frac{a-3}{a-2}\right)\cdots\left(\tfrac{3}{4}\right)\left(\tfrac{1}{2}\right)\times\\ &\quad\Big[\varphi-\sin\varphi\cos\varphi\Big]_{-\pi}^{\pi}\ \text{if } a \text{ is even}\\ &= \left(\frac{a-1}{a}\right)\left(\frac{a-3}{a-2}\right)\cdots\left(\tfrac{2}{3}\right)\Big[-\cos\varphi\Big]_{-\pi}^{\pi}\\ &\quad\text{if } a \text{ is odd}\end{aligned}\right\} \tag{61}$$

and

$$\int_{-\pi}^{\pi} \sin^a \varphi \cos \varphi \, d\varphi = \left[\frac{\sin^{a+1} \varphi}{a + 1} \right]_{-\pi}^{\pi} \tag{62}$$

If, therefore, b is even, equation (61) shows that the integral (60) will be zero unless a is even. If on the other hand b is odd, then by equation (62) the integral (60) will be zero whatever the value of a. It follows that the integral (60) will be zero unless both a and b are even, and that in this event,

$$\int_{-\pi}^{\pi} \sin^a \varphi \cos^b \varphi \, d\varphi = 2\pi \frac{[1 \,.\, 3 \,.\, 5 \ldots (a - 1)] \times [1 \,.\, 3 \,.\, 5 \ldots (b - 1)]}{2 \,.\, 4 \,.\, 6 \ldots (a + b)} \tag{63}$$

Finally, taking both a and b as even, integration of the terms containing θ with respect to θ gives

$$\left. \begin{aligned} &\int_0^{\pi/2} \sin^{a+b+1} \theta \cos^{c+1} \theta \, d\theta \\ &= \left(\frac{c}{a+b+c+2}\right)\left(\frac{c-2}{a+b+c}\right) \cdots \left(\frac{1}{a+b+3}\right) \times \\ &\quad \left(\frac{a+b}{a+b+1}\right)\left(\frac{a+b-2}{a+b-1}\right) \cdots \left(\tfrac{2}{3}\right) \text{ if } c \text{ is odd} \\ &= \left(\frac{c}{a+b+c+2}\right)\left(\frac{c-2}{a+b+c}\right) \cdots \left(\frac{2}{a+b+4}\right) \times \\ &\quad \left(\frac{1}{a+b+2}\right) \text{ if } c \text{ is even} \end{aligned} \right\}$$

from equations (60), (61) and (62), with a replaced by $a + b + 1$, b by $c + 1$, π by $\pi/2$ and $-\pi$ by 0. This result may be written

$$\left. \begin{aligned} \int_0^{\pi/2} \sin^{a+b+1} \theta \cos^{c+1} \theta \, d\theta &= \frac{[1 \,.\, 3 \,.\, 5 \ldots c] \times [2 \,.\, 4 \,.\, 6 \ldots (a+b)]}{1 \,.\, 3 \,.\, 5 \ldots (a+b+c+2)} \\ &\quad \text{if } c \text{ is odd} \\ &= \frac{2 \,.\, 4 \,.\, 6 \ldots c}{(a+b+2)(a+b+4) \ldots (a+b+c+2)} \\ &\quad \text{if } c \text{ is even} \end{aligned} \right\} \tag{64}$$

The expression (58) is now readily evaluated for various combinations of a, b and c. The sum of the results so obtained may be substituted for the triple integral in equation (55) to give, after some rearrangement,

$$\begin{aligned} J_{-z} = \tfrac{1}{4}(\Sigma_s/\Sigma_t)\Bigg[\phi &+ \frac{1/4}{\Sigma_t^2}\nabla^2\phi + \frac{1/8}{\Sigma_t^4}\nabla^4\phi + \frac{5/64}{\Sigma_t^6}\nabla^6\phi + \dots \\ &+ \left(\frac{2}{\Sigma_t}\right)\frac{\partial}{\partial z}\left\{\tfrac{1}{3}\phi + \frac{1/5}{\Sigma_t^2}\nabla^2\phi + \frac{1/7}{\Sigma_t^4}\nabla^4\phi + \dots\right\} \\ &+ \left(\frac{1/4}{\Sigma_t^2}\right)\frac{\partial^2}{\partial z^2}\left\{\phi + \frac{1}{\Sigma_t^2}\nabla^2\phi + \frac{15/16}{\Sigma_t^4}\nabla^4\phi + \dots\right\} \\ &- \left(\frac{1/8}{\Sigma_t^4}\right)\frac{\partial^4}{\partial z^4}\left\{\tfrac{1}{3}\phi + \frac{5/8}{\Sigma_t^2}\nabla^2\phi + \dots\right\} \\ &+ \left(\frac{1/64}{\Sigma_t^6}\right)\frac{\partial^6}{\partial z^6}\left\{\phi + \dots\right\} + \dots\Bigg] \end{aligned} \tag{65}$$

An expression for the current density J_{+z} may be derived in much the same way as that for the current density J_{-z}. In the case of J_{+z}, however, the coordinate z_1 of the volume element $\mathrm{d}V_1$ is invariably less than that z of the area $\mathrm{d}A$. The coordinate θ thus lies between $\pi/2$ and π so that $\cos\theta$ is invariably negative. It follows that the expression analogous to (51) for the probability that a neutron will be scattered from the volume element $\mathrm{d}V_1$ in the direction of the area $\mathrm{d}A$ must now be written

$$-\mathrm{d}A\cos\theta/4\pi r^2$$

Expressions (50) and (52) still apply, however, as also does equation (54). Finally, therefore, an expression is obtained for J_{+z} which is identical to that (55) for J_{-z} with the sign before it reversed and the limits $\pi/2$ and 0 replaced by π and $\pi/2$ respectively. Expression (56) and equations (57) apply as before, so that this replacement of limits is the only change needed in (58) when considering J_{+z}. Since

$$\int_{\pi/2}^{\pi}\sin^{a+b+1}\theta\cos^{c+1}\theta\,\mathrm{d}\theta = -(-1)^c\int_0^{\pi/2}\sin^{a+b+1}\theta\cos^{c+1}\theta\,\mathrm{d}\theta$$

it follows that J_{+z} may be set equal to an expression identical to that (65) for J_{-z} except for a change in sign of the terms for which c is odd, having odd derivatives with respect to z. The corresponding series for the nett current density J_z is therefore

$$\begin{aligned} J_z &= J_{+z} - J_{-z} \\ &= -(\Sigma_s/\Sigma_t^2)\frac{\partial}{\partial z}\left\{\tfrac{1}{3}\phi + \frac{1/5}{\Sigma_t^2}\nabla^2\phi + \frac{1/7}{\Sigma_t^4}\nabla^4\phi + \dots\right\} \end{aligned} \tag{66}$$

14.4 Infinite medium diffusion theory

Equation (66) is the relation between current density and flux that was sought for. It is here written between J_z and ϕ but similar relations between J_x or J_y and ϕ may be obtained on replacing z in equation (66) by x or y respectively. All three relations may now be substituted in equation (49) and the J thereby eliminated to give

$$\Sigma_a\phi - S = \Sigma_s\left(\frac{1/3}{\Sigma_t^{\,2}}\nabla^2\phi + \frac{1/5}{\Sigma_t^{\,4}}\nabla^4\phi + \frac{1/7}{\Sigma_t^{\,6}}\nabla^6\phi + \ldots\right) \tag{67}$$

For the case in which S is zero, a solution to this equation is sought that satisfies the relation

$$\nabla^2\phi = \phi/L^2 \tag{68}$$

where L is a constant to be determined. Substituting equation (68) in (67),

$$\Sigma_a\phi = \Sigma_s\left[\frac{1/3}{(\Sigma_t L)^2} + \frac{1/5}{(\Sigma_t L)^4} + \frac{1/7}{(\Sigma_t L)^6} + \ldots\right]\phi \tag{69}$$

$$1/\Sigma_s L = \frac{1}{(\Sigma_t L)} + \frac{1/3}{(\Sigma_t L)^3} + \frac{1/5}{(\Sigma_t L)^5} + \frac{1/7}{(\Sigma_t L)^7} + \ldots$$

which reduces to

$$1/\Sigma_s L = \tanh^{-1}(1/\Sigma_t L) \tag{70}$$

Any solution of equation (68) will thus be a solution of (67) provided that L satisfies (70). For the purely absorbing and scattering medium here considered, equations (67), (68), (69) and (70) are the appropriate forms of results reported by A. H. Wilson in 1943. The same report shows that there can be no solutions of equation (67) of a more general type than those of (68). Under certain conditions, a solution can be obtained in a similar way when the source term S is not zero. If equation (68) is modified by subtraction of the term $S/\Sigma_a L^2$ from the right-hand side, then L is given by (70) as before provided either that S is uniform or that Σ_a is very much smaller than Σ_t, whereupon all later terms of the series (69) may be ignored in comparison with the first.

With ϕ given by equation (68), (66) becomes

$$J_z = -L^2\Sigma_s\left[\frac{1/3}{(\Sigma_t L)^2} + \frac{1/5}{(\Sigma_t L)^4} + \frac{1/7}{(\Sigma_t L)^6} + \ldots\right]\partial\phi/\partial z$$

which, on introducing (69), may be written

$$\begin{aligned} J_z &= -\Sigma_a L^2\partial\phi/\partial z \\ &= -D\partial\phi/\partial z \end{aligned} \tag{71}$$

where

$$D = \Sigma_a L^2 \tag{72}$$

Equation (71) has an analogy in the process of diffusion. For the nett current density j_z of a diffusate along the coordinate z and its concentration c are connected by the relation

$$J_z = -D_c\, \partial c/\partial z$$

known as 'Fick's Law'. The quantity D_c is a constant of proportionality called the 'diffusion coefficient' and, by analogy, the constant D in equation (71) is referred to as the 'neutron diffusion coefficient'. The relation (72) enables equation (68) to be written in the form

$$D\nabla^2\phi = \Sigma_a\phi$$

where, from (48) and (71), the left-hand side represents the nett rate of neutron leakage into unit volume, and the right-hand side the rate of absorption in it. This section and the previous one form an account of 'infinite medium diffusion theory', equation (68) being termed the 'neutron diffusion equation' and the constant L appearing in it the 'diffusion length'. As noted in the previous section, the theory strictly applies only within an infinite medium but is valid up to within a few mean free paths of the boundary of a finite one.

From equations (70) and (72), $D\Sigma_t$ may be expressed in terms of Σ_a/Σ_t as shown graphically by the lower curve in Figure 14.5. As Σ_a/Σ_t becomes very small, D is seen to approach the value $\frac{1}{3}/\Sigma_t$. On introducing this value of D into equation (72), the resulting value of L^2 may be set into (68) to give the diffusion equation as used by Fermi in work published in 1936. The quantity $\frac{1}{3}/D$ is often referred to as the 'transport cross-section', Σ_{tr}.

With equation (66) for J_z reduced to the simple form (71), it remains to attempt a similar reduction of equation (65) for J_{-z}. From equations (66) and (68), (65) may be written

$$\begin{aligned} J_{\mp z} = \tfrac{1}{4}(\Sigma_s/\Sigma_t)\Bigg[\Bigg\{1 &+ \frac{1/4}{(\Sigma_t L)^2} + \frac{1/8}{(\Sigma_t L)^4} + \frac{5/64}{(\Sigma_t L)^6} + \ldots\Bigg\}\phi \\ &+ \frac{1/4}{\Sigma_t^2}\left\{1 + \frac{1}{(\Sigma_t L)^2} + \frac{15/16}{(\Sigma_t L)^4} + \ldots\right\}\frac{\partial^2\phi}{\partial z^2} \\ &- \frac{1/8}{\Sigma_t^4}\left\{\tfrac{1}{3} + \frac{5/8}{(\Sigma_t L)^2} + \ldots\right\}\frac{\partial^4\phi}{\partial z^4} + \end{aligned}$$

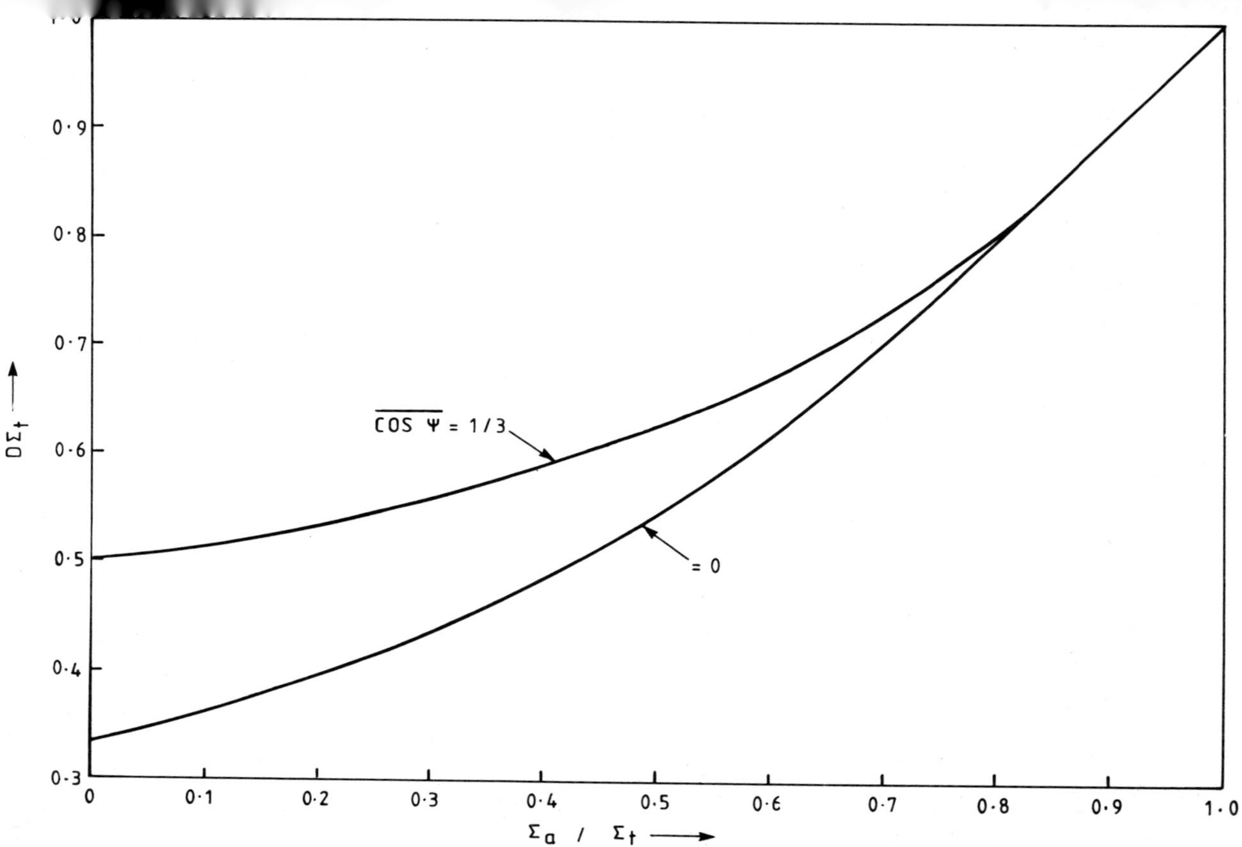

Figure 14.5 Diffusion coefficient in terms of cross-sections

$$\frac{1/64}{\Sigma_t^6}\left\{1 + \ldots\right\}\frac{\partial^6\phi}{\partial z^6} + \ldots\Bigg] \mp \tfrac{1}{2}J_z \tag{73}$$

With increasing values of Σ_a/Σ_t, it is seen from Figure 14.5 that $(\Sigma_t L)^2$, or $D\Sigma_t/(\Sigma_a/\Sigma_t)$, approaches unity so that the various infinite series in equation (73) do not converge. At the same time, equation (68) tends to the form

$$\phi = (1/\Sigma_t)^2\nabla^2\phi = (1/\Sigma_t)^4\nabla^4\phi = \ldots$$

thus providing no indication that the successive derivatives of ϕ with respect to $\Sigma_t z$ are not all comparable with ϕ.

On the other hand, when Σ_a/Σ_t is very small, $(\Sigma_t L^2)$ becomes very large so that, from equation (68),

$$(1/\Sigma_t)^2\nabla^2\phi = \phi/(\Sigma_t L)^2 << \phi$$
$$(1/\Sigma_t)^4\nabla^4\phi = (1/\Sigma_t L)^2(1/\Sigma_t)^2\nabla^2\phi = \phi/(\Sigma_t L)^4 <<<< \phi$$

and so on. It may thus be inferred that

$$(1/\Sigma_t)^2\partial^2\phi/\partial x^2,\ (1/\Sigma_t)^2\partial^2\phi/\partial y^2,\ (1/\Sigma_t)^2\partial^2\phi/\partial z^2 \text{ are each } << \phi,$$

that

$$(1/\Sigma_t)^4\partial^4\phi/\partial x^4,\ (1/\Sigma_t)^4\partial^4\phi/\partial y^4,\ (1/\Sigma_t)^4\partial^4\phi/\partial z^4 \quad \text{are each } <<<< \phi,$$

and so on. Also, as Σ_a/Σ_t vanishes, Σ_s/Σ_t approaches unity. Under these conditions, therefore, equation (73) approximates to

$$J_{\mp z} = \tfrac{1}{4}\phi \mp \tfrac{1}{2}J_z$$

and likewise (74)

$$J_{\mp x} = \tfrac{1}{4}\phi \mp \tfrac{1}{2}J_x$$
$$J_{\mp y} = \tfrac{1}{4}\phi \mp \tfrac{1}{2}J_y$$

The range of Σ_a/Σ_t over which equation (74) gives a sufficiently good approximation may be deduced from the case in which ϕ is a function of z only. In this event, equation (68) enables the several series in (73) to be combined into one, which then reduces to an analytical expression. From the result so obtained, the effect of varying Σ_a/Σ_t is readily inferred. Although straightforward, this procedure will not be adopted here since the restriction of the flux to variation with one dimension alone enables an analytical expression to be obtained for current density which accounts, not only for increased absorption but also for anisotropic scattering, as described in the following chapter.

14.5 One-dimensional scattering

In 1936, Fermi put forward the concept of 'one-dimensional scattering', which presupposed that each neutron always moved to and fro along the same straight line parallel to some fixed axis. The path of a neutron undergoing one-dimensional scattering was then regarded as the projection on that axis of the actual path under conditions of three-dimensional scattering. Although now only of historic interest, the concept was frequently used by Fermi in interpreting experimental results.

For the case of one-dimensional scattering, expression (51) may be replaced by $\frac{1}{2}$ and (54) by $d\zeta$, where $\zeta = |z_1 - z|$. The integral equation (55) then becomes, with (43),

$$J_{-z} = \tfrac{1}{2}\bar{v}'\,(\Sigma_s/\Sigma_t)\,\Sigma_t' \int_0^\infty n_1\, e^{-\Sigma_t'\zeta}\, d\zeta \qquad (75)$$

where primes denote values appropriate to one-dimensional scattering. The Taylor series given by expression (56) becomes

$$n_1 = n + (dn/dz)(z_1 - z) + \frac{1}{2!}(d^2n/dz^2)(z_1 - z)^2 + \ldots$$

so that (75) may be written

$$\begin{aligned} J_{-z} &= \tfrac{1}{2}\bar{v}'\,(\Sigma_s/\Sigma_t)\Sigma_t' \int_0^\infty \Big[n + (dn/dz)\zeta \\ &\quad + \frac{1}{2!}(d^2n/dz^2)\zeta^2 + \ldots \Big] \exp(-\Sigma_t'\,\zeta)\, d\zeta \\ &= \tfrac{1}{2}\bar{v}'\,(\Sigma_s/\Sigma_t)\left(n + \frac{dn/dz}{\Sigma_t'} + \frac{d^2n/dz^2}{\Sigma_t'^2} + \ldots\right) \end{aligned} \qquad (76)$$

from (59). In fact, Fermi limited this expression to the terms containing n and dn/dz only, but the more general form will be retained here. The current density J_{+z} is derived in a similar way but in this case

$$z_1 - z = -\zeta$$

so that the expression for J_{+z} is identical to that for J_{-z} with the exception that the signs before the odd derivatives of n are now negative. Finally, therefore, the result analogous to equation (66) becomes

$$J_z = -\bar{v}'(\Sigma_s/\Sigma_t\Sigma_t')\frac{d}{dz}\left(n + \frac{d^2n/dz^2}{\Sigma_t'^2} + \frac{d^4n/dz^4}{\Sigma_t'^4} + \ldots\right) \qquad (77)$$

For the case of one-dimensional scattering, equation (49) may be written, with $S = 0$,

$$(\Sigma_a/\Sigma_t)\,\Sigma_t'\,\overline{v}'n = -\mathrm{d}J_z/\mathrm{d}z \tag{78}$$

whilst (68) becomes

$$\mathrm{d}^2n/\mathrm{d}z^2 = n/L'^2 \tag{79}$$

On introducing equations (78) and (79) into (77), differentiated with respect to z,

$$\begin{aligned}\Sigma_a/\Sigma_t &= (\Sigma_s/\Sigma_t)\left[\frac{1}{(\Sigma_t'L')^2} + \frac{1}{(\Sigma_t'L')^4} + \dots\right] \\ &= (\Sigma_s/\Sigma_t)\,\frac{1/(\Sigma_t'L')^2}{1 - \dfrac{1}{(\Sigma_t'L')^2}}\end{aligned} \tag{80}$$

or

$$(1/L')^2 = (\Sigma_a/\Sigma_t)\,\Sigma_t'^2 \tag{81}$$

Also, equation (76) becomes, with (77), (79) and (80)

$$J_{\pm z} = \tfrac{1}{2}\overline{v}'n \pm \tfrac{1}{2}J_z \tag{82}$$

whilst (79) and (80) in (77) give

$$J_z = -(1/\Sigma_t')\,\overline{v}'\,\mathrm{d}n/\mathrm{d}z \tag{83}$$

so that, analogous to (71),

$$D' = 1/\Sigma_t' \tag{84}$$

These results for J_z, diffusion length and diffusion coefficient agree with those for three-dimensional scattering as Σ_a approaches Σ_t, provided that

$$\Sigma_t' = \Sigma_t \tag{85}$$

However, as Σ_a becomes very small relative to Σ_t, D approaches $1/(3\Sigma_t)$ so that

$$1/L^2 = 3(\Sigma_a/\Sigma_t)\,\Sigma_t^2$$

Comparison of this result with equation (81) shows that the substitution

$$\Sigma_t' = \sqrt{3}\,\Sigma_t \tag{86}$$

must now be employed in order to obtain the same diffusion equation for both one- and three-dimensional scattering. Also, in order to obtain the same reaction rate per unit volume, it is required that

$$(\Sigma/\Sigma_t)\,\Sigma_t'\,\overline{v}'n = \Sigma\overline{v}n$$

where Σ is the cross-section for the reaction considered. It follows that

$$\overline{v}' = \overline{v}/\sqrt{3}$$

The relation (86) was frequently used by Fermi in order to apply theoretical results for one-dimensional scattering to the three-dimensional case.

In the next chapter it will be seen that the distribution of angular flux at any point in an infinite medium becomes more one-dimensional as the ratio Σ_a/Σ_t increases. As might be expected, therefore, it is just under these conditions that Fermi's one-dimensional theory has been seen above to agree most closely with the three-dimensional one.

15

Angular distribution of diffusing neutrons

15.1 Anisotropic scattering

The theory derived in the previous chapter rests on the assumption that scattering is isotropic in the laboratory system. In the case of thermal neutrons this assumption is generally valid since neutrons of such low energy are scattered isotropically in the centre-of-mass system and, as follows from equation (35), in the laboratory system also except by very light atoms. It has been seen, however,that light atoms, particularly in the form of paraffin, have played a leading part in the production of thermal neutrons so that anisotropic scattering in the laboratory system is not without interest. Moreover, the infinite medium diffusion theory adopted for thermal neutrons will be of even greater value if its application can be extended to fast neutrons, the scattering of which becomes increasingly anisotropic with the mass of the scattering nucleus.

In the theory so far developed for describing the passage of neutrons through matter, the restriction to isotropic scattering is first introduced with equation (51), which expresses the probability that a neutron will be scattered within a small range of solid angle about a particular direction. In the case of anisotropic scattering, this probability will depend partly on the direction from which the neutron arrives at the scattering nucleus so that not only the flux ϕ_1 there but its angular distribution must be known. It thus becomes clear that at this stage further development of the theory must be made in terms of a quantity such as the angular flux per unit energy $\phi^{*\prime}$ rather than the flux ϕ. If this development is limited to the average behaviour of neutrons for a particular energy spectrum, however, it may more conveniently be made in terms of the 'angular flux'

$$\phi^* = \int_0^\infty \phi^{*\prime} \, \mathrm{d}E \tag{87}$$

The treatment ultimately arrived at in this way is known as 'neutron transport theory'. A discussion of it will not be given here in its most general form but will be restricted to the case in which the resulting flux ϕ varies only in one direction so that the medium considered is of infinite extent at right angles to this direction. This assumption leads to the 'one-dimensional transport theory in plane geometry'.

15.2 Neutron transport equation

Referring to Figure 15.1, the direction in which ϕ is taken to vary is parallel to the z-axis of the Cartesian coordinates x, y, z. A neutron at the point (x, y, z) travels in a direction Ω defined by the spherical coordinates θ, φ. In the one-dimensional case considered, ϕ^* is independent of x, y and φ and is a function only of z and θ. The rate at

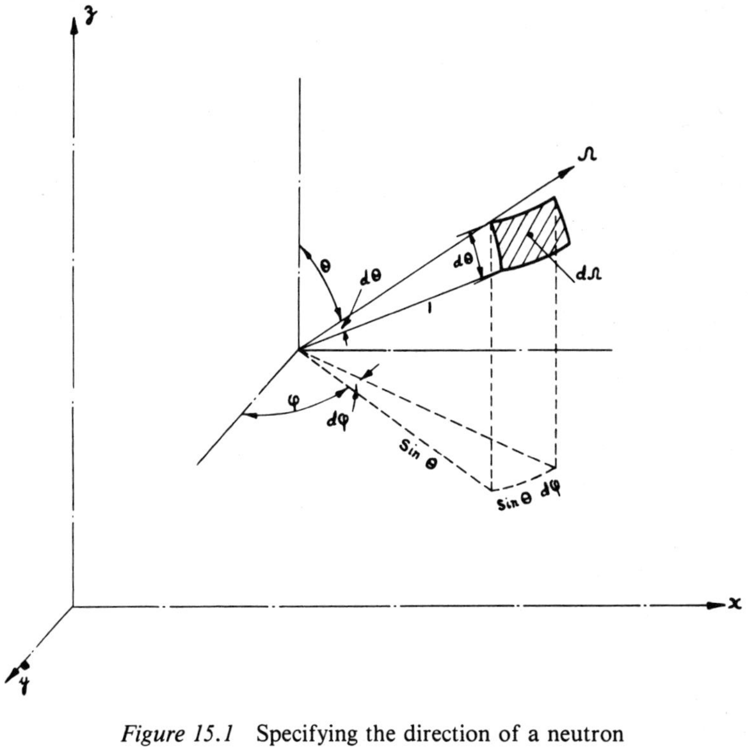

Figure 15.1 Specifying the direction of a neutron

which neutrons travelling within solid angle $\mathrm{d}\Omega$ about direction Ω cross unit area normal to that direction is

$$\phi^*(z, \theta)\,\mathrm{d}\Omega$$

and from Figure 15.1

$$\mathrm{d}\Omega = \sin\theta\,\mathrm{d}\varphi\,\mathrm{d}\theta \tag{88}$$

These same neutrons will pass through an area sec θ normal to the z-axis so that their current density in direction $+z$ is

$$\phi^*(z, \theta)\,\mathrm{d}\Omega\cos\theta \tag{89}$$

It then follows from reasoning similar to that leading up to equation (48) that the nett rate at which these neutrons leave unit volume is

$$\frac{\partial}{\partial z}\phi^*(z, \theta)\,\mathrm{d}\Omega\cos\theta \tag{90}$$

Also, integrating equation (39) with respect to E and introducing (87), the rate per unit volume at which neutrons are removed from the solid angle $\mathrm{d}\Omega$ by absorption and scattering is

$$\Sigma_t\phi^*(z, \theta)\,\mathrm{d}\Omega \tag{91}$$

Finally, an expression must be obtained for the rate per unit volume at which neutrons are scattered into the solid angle $\mathrm{d}\Omega$ about direction Ω from all directions. From the definition of the differential scattering cross-section σ_ψ given in Section 12.4, it follows that the cross-section for scattering into the solid angle $\mathrm{d}\Omega$ through the angle ψ is $\Sigma_\psi\,\mathrm{d}\Omega$. Hence the rate per unit volume at which neutrons are scattered into this solid angle from the solid angle $\mathrm{d}\Omega'$ about direction Ω', making an angle ψ with direction Ω, is

$$\Sigma_\psi\,\mathrm{d}\Omega\,\phi^*(z, \theta')\,\mathrm{d}\Omega' = \Sigma_\psi\,\mathrm{d}\Omega\,\phi^*(z, \theta')\sin\theta'\,\mathrm{d}\theta'\,\mathrm{d}\varphi'$$

from equation (88). The required expression for the corresponding rate of scattering into the solid angle $\mathrm{d}\Omega$ from all directions is thus

$$\mathrm{d}\Omega\int_0^\pi\int_{-\pi}^\pi \Sigma_\psi\phi^*(z, \theta')\sin\theta'\,\mathrm{d}\varphi'\,\mathrm{d}\theta' \tag{92}$$

Since this expression must be equal to the sum of expressions (90) and (91), there results the equation

$$\cos\theta\frac{\partial}{\partial z}\phi^*(z,\theta) + \Sigma_t\phi^*(z,\theta) = \int_0^\pi\int_{-\pi}^\pi \Sigma_\psi\phi^*(z,\theta')\sin\theta'\,\mathrm{d}\varphi'\,\mathrm{d}\theta' \tag{93}$$

which is the 'one-dimensional transport equation in plane geometry'.

15.3 Differential scattering cross-section in spherical coordinates

In order to evaluate the integral in this equation, the differential scattering cross-section must be expressed as a function of the spherical coordinates θ, φ and θ', φ' rather than the scattering angle ψ. The relation between these coordinates and ψ may be found with the aid of Figure 15.2. The direction Ω, having the coordinates θ, φ, is represented by a line joining the origin O of a sphere of unit radius to a point A on its surface. Likewise, the direction Ω', having the coordinates θ', φ', is represented by a line OB. The plane through B at right angles to the diameter $\theta = 0$ cuts this diameter at C and the line parallel to it through A at D. Then

$$CD = \sin\theta$$

and

$$BC = \sin\theta'$$

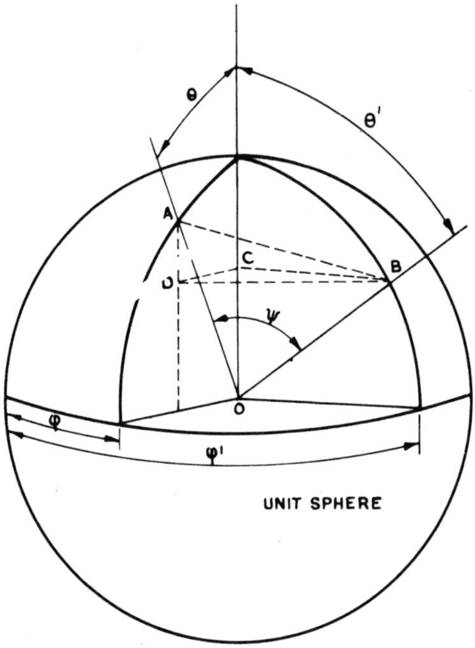

Figure 15.2 Scattering angle in terms of spherical coordinates

so that

$$BD^2 = \sin^2\theta + \sin^2\theta' - 2\sin\theta\sin\theta'\cos(\varphi' - \varphi)$$

Also

$$AD = \cos\theta - \cos\theta'$$

so that

$$\begin{aligned} AB^2 &= AD^2 + BD^2 \\ &= \cos^2\theta - 2\cos\theta\cos\theta' + \cos^2\theta' + \sin^2\theta + \sin^2\theta' \\ &\quad - 2\sin\theta\sin\theta'\cos(\varphi' - \varphi) \end{aligned}$$

But

$$\begin{aligned} AB^2 &= OA^2 + OB^2 - 2OA\,.\,OB\cos\psi \\ &= 2(1 - \cos\psi) \end{aligned}$$

Finally, therefore,

$$\cos\psi = \cos\theta\cos\theta' + \sin\theta\sin\theta'\cos(\varphi' - \varphi) \quad (94)$$

As a first approximation, it is convenient to set

$$\Sigma_\psi = C_0 + 3C_1\cos\psi \quad (95)$$

The significance of the coefficients C_0, C_1 may be examined in the following way. From Figure 12.4, the solid angle corresponding to the increment $d\psi$ of the scattering angle ψ is seen to be $2\pi\sin\psi\, d\psi$. The cross-section for scattering through an angle between ψ and $\psi + d\psi$ is therefore

$$\Sigma_\psi\,.\,2\pi\sin\psi\, d\psi$$

so that

$$\left.\begin{aligned} \Sigma_s &= 2\pi\int_0^\pi \Sigma_\psi \sin\psi\, d\psi \\ \Sigma_s\overline{\cos\psi} &= 2\pi\int_0^\pi \Sigma_\psi \cos\psi\sin\psi\, d\psi \end{aligned}\right\} \quad (96)$$

where $\overline{\cos\psi}$ is the mean value of $\cos\psi$, known as the 'mean cosine of the scattering angle'. From equation (95), the last two results become

$$\begin{aligned} \Sigma_s &= 4\pi C_0 \\ \Sigma_s\overline{\cos\psi} &= 4\pi C_1 \end{aligned}$$

and (95) may be written

$$\Sigma_\psi = (\Sigma_s/4\pi)(1 + 3\,\overline{\cos\psi}\,.\,\cos\psi) \tag{97}$$

On introducing equations (94) and (97) into (93), the integration with respect to φ' may be carried out to give

$$\cos\theta\,\frac{\partial}{\partial z}\,\phi^*(z,\theta) + \Sigma_t\,\phi^*(z,\theta)$$
$$= \tfrac{1}{2}\Sigma_s\int_0^\pi (1 + 3\,\overline{\cos\psi}\cos\theta\cos\theta')\,\phi^*(z,\theta')\sin\theta'\,d\theta' \tag{98}$$

From equations (33) and (34), the second of equations (96) may be written

$$\overline{\cos\psi} = (2\pi/\Sigma_s)\int_{-1}^{1}\Sigma_\theta\,\frac{1 + A\cos\theta}{\sqrt{(1 + A^2 + 2A\cos\theta)}}\,d(\cos\theta) \tag{99}$$

Since σ_θ is generally presented as a plot against $\cos\theta$, this result enables the corresponding value of $\overline{\cos\psi}$ to be readily determined by graphical or numerical integration. In the case of isotropic scattering in the centre-of-mass system, Σ_θ is seen from equation (30) to equal $\Sigma_s/4\pi$ so that the integration may be performed analytically to give

$$\overline{\cos\psi} = (1/6A)\Big[(2 - A^2 + A\cos\theta)\sqrt{(1 + A^2 + 2A\cos\theta)}\Big]_{-1}^{1}$$
$$= 2/3A \tag{100}$$

From this result, $\overline{\cos\psi}$ is seen to reach a maximum value of 2/3 when A is unity. When scattering is not isotropic in the centre-of-mass system, however, higher values of $\overline{\cos\psi}$ are possible. For instance, if

$$\left.\begin{aligned}\Sigma_\psi &= k_0 \qquad & 0 \le \psi < \psi_0\\ &= 0 & \psi_0 \le \psi \le \pi\end{aligned}\right\}$$

then, from equations (96),

$$\begin{aligned}\Sigma_s &= 2\pi k_0(1 - \cos\psi_0)\\ \Sigma_s\,\overline{\cos\psi} &= \pi k_0(1 - \cos^2\psi_0)\end{aligned}$$

whence

$$\overline{\cos\psi} = \tfrac{1}{2}(1 + \cos\psi_0)$$

It follows that $\overline{\cos\psi}$ tends to unity as ψ_0 approaches zero, the value of k_0 then becoming infinite if Σ_s is to be other than zero. However, use of the linear relationship (97) between Σ_ψ and $\cos\psi$ sets an upper limit of 1/3 to $\overline{\cos\psi}$ if Σ_ψ is not to become negative.

15.4 Solution by separation of variables

To find a solution to the transport equation (93) an attempt may be made by separating the variables z and θ so that the trial solution takes the form

$$\phi^*(z, \theta) = F(z) f(\theta) \tag{101}$$

where $F(z)$ is a function of z only and $f(\theta)$ a function of θ only. From the identity of equations (43) with (42) and of (41) with (40), it follows that

$$\begin{aligned} \phi &= \int_0^{4\pi} \int_0^{\infty} \phi^{*\prime}(\Omega, E) \, dE \, d\Omega \\ &= \int_0^{\pi} \int_{-\pi}^{\pi} \phi^*(z, \theta) \sin\theta \, d\varphi \, d\theta \end{aligned} \tag{102}$$

from (87) and (88). On carrying out the integration with respect to φ and introducing equation (101),

$$\phi(z) = 2\pi F(z) \int_0^{\pi} f(\theta) \sin\theta \, d\theta$$

so that (101) may be written

$$\phi^*(z, \theta) = \phi(z) f(\theta) / 2\pi \int_0^{\pi} f(\theta') \sin\theta' \, d\theta' \tag{103}$$

With this relation, equation (98) becomes, on dividing through by $\phi(z) f(\theta) \cos\theta$,

$$\begin{aligned} \frac{1}{\phi(z)} \cdot \frac{d}{dz} \phi(z) = {} & \frac{1}{\cos\theta} \left[\frac{\frac{1}{2}\Sigma_s}{f(\theta)} \int_0^{\pi} f(\theta') \sin\theta' \, d\theta' - \Sigma_t \right] \\ & + \frac{3\Sigma_s}{2f(\theta)} \overline{\cos\psi} \int_0^{\pi} f(\theta') \cos\theta' \sin\theta' \, d\theta' \end{aligned} \tag{104}$$

Since the left-hand side of this equation is a function of z only and the right-hand side is a function of θ only, and since z and θ may be varied independently, each side of the equation must be equal to one and the same constant, which will be written $\pm \varkappa$. Setting the left-hand side of equation (104) equal to this constant,

$$\frac{d}{dz} \phi(z) = \pm \varkappa \, \phi(z) \tag{105}$$

$$\frac{d^2}{dz^2}\,\phi(z) = \pm\varkappa\,\frac{d}{dz}\,\phi(z)$$
$$= \varkappa^2\,\phi(z) \tag{106}$$

Comparison of this result with equation (68) shows that, when $\overline{\cos\psi}$ is zero, the constant $\varkappa$ may be identified with the inverse diffusion length $1/L$. Henceforward, however, L will be regarded as a function of $\overline{\cos\psi}$ so that $\varkappa$ may be identified with $1/L$ whether $\overline{\cos\psi}$ is zero or not. Corresponding to the values $+1/L$, $-1/L$ for $\sqrt{\varkappa^2}$, the respective solutions of equation (106) will be written

$$\phi_1(z),\ \phi_2(z) \tag{107}$$

It remains to derive an expression for L and the form of the function $f(\theta)$. It is clear that equation (104) remains unaffected if $f(\theta)$ is multiplied by any one and the same constant factor wherever it occurs and the form of the function $f(\theta)$ is therefore arbitrary to the extent of such a factor. This factor will now be arbitrarily chosen so that

$$\tfrac{1}{2}\Sigma_s L\int_0^\pi f(\theta')\sin\theta'\,d\theta' = 1 \tag{108}$$

whereupon

$$\tfrac{3}{2}\Sigma_s L\,\overline{\cos\psi}\int_0^\pi f(\theta')\cos\theta'\,\sin\theta'\,d\theta' = a \tag{109}$$

where a is a constant to be determined. With these two substitutions, the right-hand side of equation (104) may be set equal to $\pm 1/L$ to give

$$f_{1,2}(\theta) = (1 + a\cos\theta)/(\Sigma_t L \pm \cos\theta) \tag{110}$$

where, as in (107), suffix 1 is appropriate to $+1/L$ and suffix 2 to $-1/L$. Introducing equation (110) into (108)

$$1 = \tfrac{1}{2}\Sigma_s L\Big[(a\Sigma_t L \mp 1)\log(\Sigma_t L \pm \cos\theta') \mp a\cos\theta'\Big]_0^\pi$$

$$= \mp\tfrac{1}{2}\Sigma_s L\{(a\Sigma_t L \mp 1)\log[(\Sigma_t L + 1)/(\Sigma_t L - 1)] - 2a\} \tag{111}$$

Similarly, introducing equation (110) into (109),

$$a = \tfrac{3}{2}\Sigma_s L\,\overline{\cos\psi}\Big[(a\Sigma_t L \mp 1)\cos\theta' \mp$$
$$\Sigma_t L(a\Sigma_t L \mp 1)\log(\Sigma_t L \pm \cos\theta') \mp \tfrac{1}{2}a\cos^2\theta'\Big]_0^\pi \tag{112}$$

$$= \tfrac{3}{2}\Sigma_s L\,\overline{\cos\psi}\,(a\Sigma_t L \mp 1)\{\Sigma_t L \log[(\Sigma_t L + 1)/(\Sigma_t L - 1)] - 2\}$$

Eliminating the logarithm in this equation by means of (111),

$$a = \mp 3\Sigma_a L\,\overline{\cos\psi} \tag{113}$$

whence (111) becomes

$$1 + 3\Sigma_a\Sigma_s L^2\,\overline{\cos\psi} = \tfrac{1}{2}\Sigma_s L(1 + 3\Sigma_a\Sigma_t L^2\,\overline{\cos\psi}) \times \log[(\Sigma_t L + 1)/(\Sigma_t L - 1)] \tag{114}$$

or

$$\Sigma_s L \tanh^{-1}(1/\Sigma_t L) = (1 + 3\Sigma_a\Sigma_s L^2\,\overline{\cos\psi})/(1 + 3\Sigma_a\Sigma_t L^2\,\overline{\cos\psi}) \tag{115}$$

This equation is the one giving L that was sought for, and is the form appropriate to the purely absorbing and scattering medium here considered of a result given by A. H. Wilson in 1944. From equations (113) and (110), the required form of the function $f(\theta)$ is found to be

$$f_{1,2}(\theta) = (1 \mp 3\Sigma_a L\,\overline{\cos\psi}\cos\theta)/(\Sigma_t L \pm \cos\theta) \tag{116}$$

whilst (108) into (103) gives

$$\phi^*_{1,2}(z,\theta) = \frac{\Sigma_s L}{4\pi}\phi_{1,2}(z)f_{1,2}(\theta) \tag{117}$$

Equation (98) is satisfied by the sum of any solutions for $\phi^*(z,\theta)$ that satisfy it individually so that, finally,

$$\phi^*(z,\theta) = A_1\phi_1^*(z,\theta) + A_2\phi_2^*(z,\theta) \tag{118}$$

where A_1, A_2 are arbitrary constants. Substitution of this result into equation (102) gives

$$\phi(z) = A_1\phi_1(z) + A_2\phi_2(z) \tag{119}$$

When $\overline{\cos\psi}$ is zero, equation (115) reduces, as it should, to (70) whilst (116) gives

$$f_{1,2}(\theta) = 1/(\Sigma_t L \pm \cos\theta)$$

When Σ_a is very small, Σ_s approaches Σ_t so that the right-hand side of equation (115) becomes very nearly unity. It follows that $\Sigma_t L$ is very large and equation (115) may thus be written

$$\Sigma_s L\left[\frac{1}{\Sigma_t L} + \frac{1/3}{(\Sigma_t L)^3} + \frac{1/5}{(\Sigma_t L)^5} + \ldots\right](1 + 3\Sigma_a\Sigma_t L^2\,\overline{\cos\psi})$$
$$= 1 + 3\Sigma_a\Sigma_s L^2\,\overline{\cos\psi}$$

or, ignoring terms of order less than $(1/\Sigma_t L)^4$, $(\Sigma_a/\Sigma_t)^2$ or $(1/\Sigma_t L)^2(\Sigma_a/\Sigma_t)$,

$$(1/\Sigma_t L)^2\{\tfrac{1}{3}[1 - (\Sigma_a/\Sigma_t)] + \tfrac{3}{5}(\Sigma_a/\Sigma_t)\overline{\cos\psi}\} + \tfrac{1}{5}(1/\Sigma_t L)^4$$
$$= (\Sigma_a/\Sigma_t)\{1 - [1 - (\Sigma_a/\Sigma_t)]\overline{\cos\psi}\}$$

As a first approximation, terms of order less than $(1/\Sigma_t L)^2$ or Σ_a/Σ_t may also be ignored to give

$$1/3\Sigma_a\Sigma_t L^2 = 1 - \overline{\cos\psi} \tag{120}$$

Multiplying this relation through by $(1/\Sigma_t L)^2$ and then introducing it into the $(1/\Sigma_t L)^4$ term, a second approximation is obtained in the form

$$1/(3\Sigma_a\Sigma_t L^2) = (1 - \overline{\cos\psi}) - \tfrac{1}{5}(4 - 9\,\overline{\cos\psi})(\Sigma_a/\Sigma_t)$$

Equation (116) may be put in the form

$$f_{1,2}(\theta) = (1/\Sigma_t L)(1 \mp 3\Sigma_a L\,\overline{\cos\psi}\cos\theta)[1 \pm (\cos\theta/\Sigma_t L)]^{-1}$$
$$= (1/\Sigma_t L)\left[1 + (3\Sigma_a\Sigma_t L^2\,\overline{\cos\psi} + 1)\left\{\mp\frac{\cos\theta}{\Sigma_t L} + \left(\frac{\cos\theta}{\Sigma_t L}\right)^2 \mp \left(\frac{\cos\theta}{\Sigma_t L}\right)^3 + \ldots\right\}\right]$$

which, on introducing the immediately previous relation for $3\Sigma_a\Sigma_t L^2$ and ignoring terms of order $(\Sigma_a/\Sigma_t)^2$ or less, reduces to

$$f_{1,2}(\theta) = (1/\Sigma_t L)[1 \mp 3\Sigma_a L\{1 - \tfrac{1}{5}(4 - 9\,\overline{\cos\psi})(\Sigma_a/\Sigma_t)\}\cos\theta$$
$$+ 3(\Sigma_a/\Sigma_t)\cos^2\theta \mp 9\Sigma_a L(\Sigma_a/\Sigma_t)(1 - \overline{\cos\psi})\cos^3\theta] \tag{121}$$

From this result, as absorption becomes very small, the angular distribution of neutrons is seen to approach one proportional to $\cos\theta$ superimposed upon a uniform one.

As already mentioned above, use of the approximation (97) sets an upper limit to $\overline{\cos\psi}$ of 1/3. More generally, however, the expression (97) for Σ_ψ may be expanded to include additional terms containing successively higher powers of $\cos\psi$, so that the rather artificial upper limit of 1/3 to $\overline{\cos\psi}$ may be removed. The diffusion length L is then given, for sufficiently small absorption, by the more general result

$$1/3\Sigma_a\Sigma_t L^2 = (1 - \overline{\cos\psi}) - \left[\frac{(4/5)(1 - \overline{\cos\psi})}{1 - \tfrac{1}{2}(3\,\overline{\cos^2\psi} - 1)} - \overline{\cos\psi}\right]\frac{\Sigma_a}{\Sigma_t}$$

where terms of order less than Σ_a/Σ_t on the right-hand side are ignored and where $\overline{\cos^2\psi}$ is obtained from equations (96) with $\cos\psi$

replaced by $\cos^2 \psi$. This result has been derived by means which are beyond the scope of the present treatment. However, it is seen to take the form of the second approximation immediately following (120) when $\overline{\cos^2 \psi}$ is equal to 1/3, which it will be when Σ_ψ conforms to the approximation (97). When absorption becomes even smaller so that the second term on the right-hand side can be ignored in comparison with the first, then the more general result reduces to the approximation (120), which is then valid whatever the value of $\overline{\cos \psi}$. As absorption increases, however, results for L obtained from equation (115) will become more strongly dependent on the strictly linear relationship between Σ_ψ and $\cos \psi$ on which that equation is based. The upper limit of 1/3 on $\overline{\cos \psi}$ must then be observed.

15.5 Current densities

Turning next to considerations of current density, expression (89) gives, with (88),

$$J_{+z} = \int_0^{\pi/2}\int_{-\pi}^{\pi} \phi^*(z,\theta) \cos\theta \sin\theta \, d\varphi \, d\theta \tag{122}$$

or

$$(J_{+z})_{1,2} = \tfrac{1}{2}\Sigma_s L \, \phi_{1,2}(z) \int_0^{\pi/2} f_{1,2}(\theta) \cos\theta \sin\theta \, d\theta$$

from (117), where the suffixes have the same significance as in the previous section. The integration may be performed by equating the left-hand side of equation (109) to the right-hand side of (112) and substituting the limit $\pi/2$ for π. It follows that

$$\begin{aligned}(J_{+z})_{1,2} &= \tfrac{1}{2}\Sigma_s L \, \phi_{1,2}(z)\{-(a\Sigma_t L \mp 1) \mp \\ &\quad \Sigma_t L(a\Sigma_t L \mp 1) \log [\Sigma_t L/(\Sigma_t L \pm 1)] \pm \tfrac{1}{2}a\} \\ &= \tfrac{1}{2}\Sigma_s L \, \phi_{1,2}(z)\left[(1 + 3\Sigma_a\Sigma_t L^2 \overline{\cos \psi})\left(\Sigma_t L \log \frac{\Sigma_t L}{\Sigma_t L \pm 1}\right.\right. \\ &\quad \left.\left. \pm 1\right) - (3/2)\Sigma_a L \overline{\cos \psi}\right]\end{aligned} \tag{123}$$

from (113). Again,

$$J_{-z} = -\int_{\pi/2}^{\pi}\int_{-\pi}^{\pi} \phi^*(z, \theta) \cos\theta \sin\theta \, d\varphi \, d\theta \tag{124}$$

an expression identical to that (122) for J_{+z} with the limit 0 replaced by π. If the change in limit is observed, therefore, this expression may be subjected to the same procedure as was used to derive equation (123) from (122), so that

$$(J_{-z})_{1,2} = \tfrac{1}{2}\Sigma_s L\,\phi_{1,2}(z)\left[(1 + 3\Sigma_a\Sigma_t L^2\,\overline{\cos\psi})\left(\Sigma_t L \log\frac{\Sigma_t L}{\Sigma_t L \pm 1} \pm 1\right) - (3/2)\Sigma_a L\,\overline{\cos\psi}\right] \tag{125}$$

Also, subtracting equation (125) from (123),

$$(J_z)_{1,2} = \mp\tfrac{1}{2}\Sigma_s L\,\phi_{1,2}(z)(1 + 3\Sigma_a\Sigma_t L^2\,\overline{\cos\psi}) \times \{\Sigma_t L \log[(\Sigma_t L + 1)/(\Sigma_t L - 1)] - 2\} \tag{126}$$

$$= \mp\Sigma_a L\,\phi_{1,2}(z) \tag{127}$$

on introducing (114). Finally,

$$J_z = A_1 J_{z1} + A_2 J_{z2} \tag{128}$$

$$= -\Sigma_a L(A_1\phi_1 - A_2\phi_2)$$

which, with equations (105) and (119), may be written

$$J_z = -\Sigma_a L^2 \frac{\mathrm{d}}{\mathrm{d}z}\phi(z)$$

$$= -D\frac{\mathrm{d}}{\mathrm{d}z}\phi(z) \tag{129}$$

analogous to (71). With the aid of equation (115) $D\Sigma_t$ may be plotted against Σ_a/Σ_t for various values of $\overline{\cos\psi}$. Two cases are shown in Figure 14.5. As absorption becomes very small, the curves are seen to agree with equation (120) which may be rewritten in the form $D\Sigma_t = \dfrac{1/3}{1 - \overline{\cos\psi}}$. The curve for $\overline{\cos\psi} = 0$ is that already obtained by diffusion theory whilst the curve for $\overline{\cos\psi} = 1/3$ corresponds to the limit set by the use of the relation (97).

Introducing equation (126) into (123),

$$(J_{+z})_{1,2} = \tfrac{1}{2}\Sigma_s L\,\phi_{1,2}(z)\left[(1 + 3\Sigma_a\Sigma_t L^2\,\overline{\cos\psi})\,\Sigma_t L\left(\log\frac{\Sigma_t L}{\Sigma_t L \pm 1} \pm \tfrac{1}{2}\log\frac{\Sigma_t L + 1}{\Sigma_t L - 1}\right) - (3/2)\,\Sigma_a L\,\overline{\cos\psi}\right] + \tfrac{1}{2}(J_z)_{1,2}$$

which, with the identity

$$\log[\Sigma_t L/(\Sigma_t L \pm 1)] \pm \tfrac{1}{2}\log[(\Sigma_t L + 1)/(\Sigma_t L - 1)]$$

$$= \tfrac{1}{2}\log\left[\left(\frac{\Sigma_t L}{\Sigma_t L \pm 1}\right)^2 \frac{\Sigma_t L \pm 1}{\Sigma_t L \mp 1}\right]$$

becomes

$$J_{+z} = \tfrac{1}{4}c_z\phi(z) + \tfrac{1}{2}J_z \tag{130}$$

where

$$c_z = (1 + 3\Sigma_a\Sigma_t L^2 \overline{\cos\psi})(\Sigma_s/\Sigma_t)(\Sigma_t L)^2 \log\left[\frac{1}{1 - (1/\Sigma_t L)^2}\right] - 3\Sigma_a\Sigma_s L^2 \overline{\cos\psi} \tag{131}$$

Similarly, equation (126) into (125) gives

$$J_{-z} = \tfrac{1}{4}c_z\phi(z) - \tfrac{1}{2}J_z \tag{132}$$

The function c_z is plotted against Σ_a/Σ_t in Figure 15.3 for the same two values of $\overline{\cos\psi}$ as before.

In order to find the current densities in the x- and y-directions, it is first noted from Figure 15.1 that neutrons travelling within solid angle $d\Omega$ about direction Ω that cross unit area normal to that direction will pass through an area cosec θ cosec φ normal to the x-axis. Their current density in the direction $+x$ is, then, analogous to expression (89),

$$\phi^*(z, \theta)\, d\Omega \sin\theta \sin\varphi$$

so that, with equation (88),

$$J_{+x} = \int_0^\pi \int_0^\pi \phi^*(z, \theta) \sin^2\theta \sin\varphi \, d\varphi\, d\theta$$

and

$$(J_{+x})_{1,2} = (\Sigma_s L/2\pi)\, \phi_{1,2}(z) \int_0^\pi f_{1,2}(\theta) \sin^2\theta\, d\theta \tag{133}$$

from (117). With $f_{1,2}(\theta)$ as given by equation (110),

$$\int_0^\pi f_{1,2}(\theta) \sin^2\theta\, d\theta = \pm \Big[(a\Sigma_t L \mp 1)\{\pm\sin\theta - (\Sigma_t L)\theta \mp \sqrt{[(\Sigma_t L)^2 - 1]}\, \sin^{-1}[(1 \pm \Sigma_t L\cos\theta)/(\Sigma_t L \pm \cos\theta)]\} - \tfrac{1}{2}a(\cos\theta\sin\theta - \theta) \Big]_0^\pi$$

$$= \mp\pi[(a\Sigma_t L \mp 1)\{\Sigma_t L - \sqrt{[(\Sigma_t L)^2 - 1]}\} - \tfrac{1}{2}a]$$

From this result and equation (113), (133) becomes

$$J_{+x} = \tfrac{1}{4}c_x\phi(z) \tag{134}$$

where

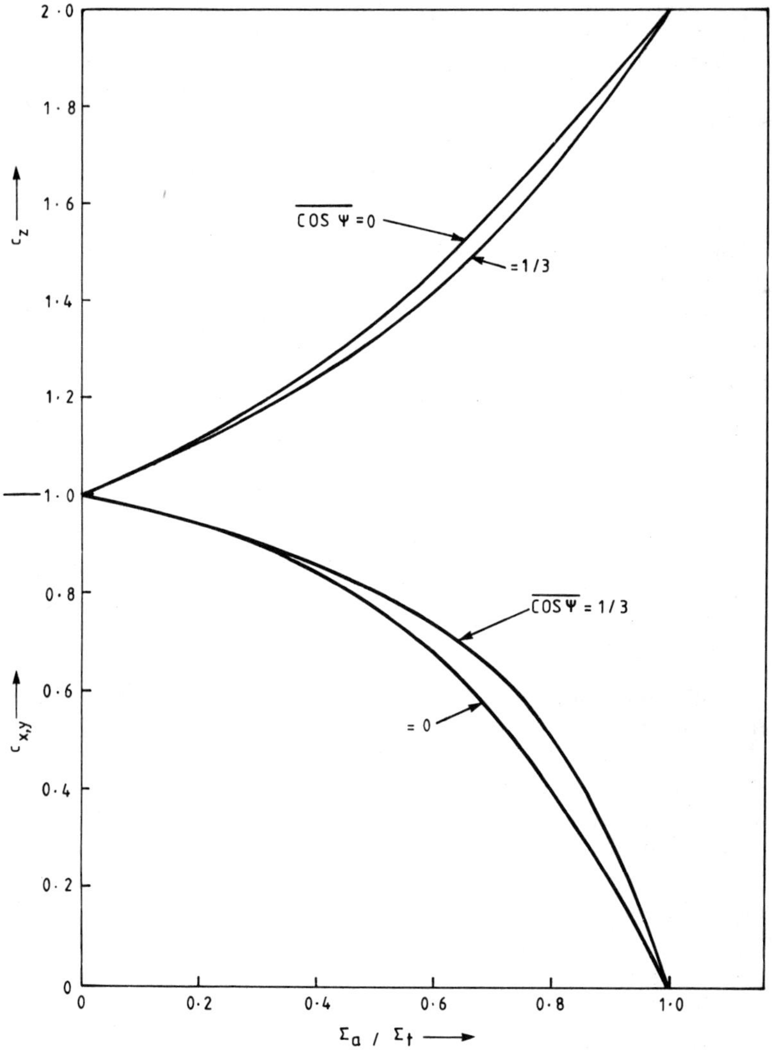

Figure 15.3 The functions c_z and $c_{x,y}$

$$c_x = \Sigma_s L\,[2(1 + 3\Sigma_a\Sigma_t L^2\,\overline{\cos\psi})\{\Sigma_t L - \sqrt{[(\Sigma_t L)^2 - 1]}\} - 3\Sigma_a L\,\overline{\cos\psi}] \tag{135}$$

Since $\phi^*(z, \theta)$ is independent of φ, it is clear that

$$J_{+x} = J_{-x} = J_{+y} = J_{-y} \tag{136}$$

so that

$$J_x = J_y = 0$$

and

$$c_x = c_y$$

being written $c_{x,y}$. The function $c_{x,y}$ is also shown plotted against Σ_a/Σ_t in Figure 15.3.

If absorption is very small, c_z and $c_{x,y}$ approach unity so that equations (130), (132), (134) and (136) reduce to

$$\left.\begin{aligned} J_{\pm z} &= \tfrac{1}{4}\phi(z) \pm \tfrac{1}{2}J_z \\ J_{\pm x} &= \tfrac{1}{4}\phi(z) \\ J_{\pm y} &= \tfrac{1}{4}\phi(z) \end{aligned}\right\} \tag{137}$$

Under these conditions, the angular distribution of neutrons is given by equation (121) and approximates to the constant and cos θ terms only. If J_z is now set equal to zero, so also must be the cos θ term of the angular distribution which thus reduces to a uniform one. It follows that the term $\frac{1}{4}\phi$ in equations (137) is the current density due to a uniform angular distribution of neutrons.

It may be noted that equations (106) and (129) of one-dimensional transport theory are identical to the corresponding equations (68) and (71) of three-dimensional infinite medium diffusion theory as applied to the one-dimensional case. Furthermore, for the case of isotropic scattering to which diffusion theory is restricted, equation (115) of transport theory becomes identical to (70) of diffusion theory to give the same values of D or L. It thus appears that the above solution to the transport equation (93) is valid, like diffusion theory, only in an infinite medium or, less restrictively, in a medium of infinite extent in the x- and y-directions and to within a few mean free paths of the boundary in the z-direction. It is therefore referred to as the 'infinite medium solution'. The spatial limitations to which it is subject are not inherent in the transport equation itself, however, and a solution that is free of them will be briefly discussed in the next chapter. For the present, the infinite medium solution not only provides information on the angular distribution of neutrons but also enables the diffusion theory expressions for D and L to be modified for the effects of anisotropic scattering. Although this modification has only been justified for the one-dimensional case, it will also be made in the three-dimensional one.

When absorption is very small, three-dimensional infinite medium diffusion theory enables the various current densities to be found from equations (74). As applied to the one-dimensional case in plane geometry these become identical to equations (137) of infinite medium transport theory, which are again restricted to the case of small absorption. With increased absorption an indication of the error involved in the use of equations (74) can readily be obtained for the one-dimensional case in plane geometry from (130), (132), (134) and Figure 15.3. It is then seen that if Σ_a/Σ_t is less than 0.1, the error in taking c_z, $c_{x,y}$ equal to unity, as in equations (74), is less than 5 per cent in the case of the first term of $J_{\pm z}$ and less than $2\frac{1}{2}$ per cent in the case of the other current densities. As may be seen from Section 22.3 however, when the medium is the core of a nuclear reactor the coefficients c_z, $c_{x,y}$ are much less sensitive to the cross-sections, so that equations (74) become more generally valid.

15.6 Distribution of angular flux

The distribution of angular flux is given by equations (116), (117) and (118) as

$$\phi^*(z, \theta) = (\Sigma_s L/4\pi)\left[A_1\phi_1 \frac{1 - 3\Sigma_a L\,\overline{\cos\psi}\cos\theta}{\Sigma_t L + \cos\theta} + A_2\phi_2 \frac{1 + 3\Sigma_a L\,\overline{\cos\psi}\cos\theta}{\Sigma_t L - \cos\theta}\right]$$

Introducing equations (119), (127) and (128), this result may be written

$$\phi^*(z, \theta) = \frac{(1/4\pi)(\Sigma_s/\Sigma_a)}{(\Sigma_t L)^2 - \cos^2\theta}\{\phi(\Sigma_a\Sigma_t L^2)[1 + 3(\Sigma_a/\Sigma_t)\,\overline{\cos\psi}\cos^2\theta] + J_z(1 + 3\Sigma_a\Sigma_t L^2\,\overline{\cos\psi})\cos\theta\} \tag{138}$$

When terms of order $(\Sigma_a/\Sigma_t)^2$ or less may be ignored, equations (121), (117) and (118) give

$$\phi^*(z, \theta) = \frac{\Sigma_s/\Sigma_t}{4\pi}\left[(A_1\phi_1 + A_2\phi_2)\{1 + 3(\Sigma_a/\Sigma_t)\cos^2\theta\} - 3\Sigma_a L(A_1\phi_1 - A_2\phi_2)\left\{\left(1 - \frac{4 - 9\,\overline{\cos\psi}}{5}\frac{\Sigma_a}{\Sigma_t}\right)\cos\theta + 3(\Sigma_a/\Sigma_t)(1 - \overline{\cos\psi})\cos^3\theta\right\}\right]$$

which, with (119), (127) and (128), becomes

$$\phi^*(z, \theta) = \frac{1}{4\pi}\,[\phi\{1 + (\Sigma_a/\Sigma_t)(3\cos^2\theta - 1)\} + 3J_z\{\cos\theta + 3(\Sigma_a/\Sigma_t)(1 - \overline{\cos\psi})(\cos^3\theta - \tfrac{3}{5}\cos\theta)\}] \tag{139}$$

If absorption is even smaller, so that terms of order Σ_a/Σ_t may be ignored, then, as noted in the previous section, the angular flux becomes the sum of two components, a constant term proportional to the flux, and a $\cos\theta$ term proportional to the current density.

At the other extreme when Σ_a approaches Σ_t, $\Sigma_t L$ is very nearly unity and equation (138) reduces to

$$\phi^*(z, \theta) = \frac{(1/4\pi)(\Sigma_s/\Sigma_t)}{1 - \cos^2\theta}\,[\phi(1 + 3\,\overline{\cos\psi}\cos^2\theta) + J_z(1 + 3\,\overline{\cos\psi})\cos\theta] \tag{140}$$

unless $\cos\theta$ approaches ± 1, when $1 - \cos^2\theta$ must be replaced by $(\Sigma_t L)^2 - \cos^2\theta$ where, from (115),

$$\Sigma_t L = \coth\frac{\Sigma_t/\Sigma_s}{1 + 3\,\overline{\cos\psi}}.$$

When $\cos\theta$ is equal to ± 1, therefore,

$$(\Sigma_t L)^2 - \cos^2\theta \approx 4\exp[-2(\Sigma_t/\Sigma_s)/(1 + 3\,\overline{\cos\psi})]$$

so that

$$\frac{\Sigma_s/\Sigma_t}{(\Sigma_t L)^2 - \cos^2\theta} \approx \frac{1}{4}\left[(\Sigma_s/\Sigma_t) + \frac{2}{1 + 3\,\overline{\cos\psi}} + (4/2!)\frac{\Sigma_t/\Sigma_s}{(1 + 3\,\overline{\cos\psi})^2} + (8/3!)\frac{(\Sigma_t/\Sigma_s)^2}{(1 + 3\,\overline{\cos\psi})^3} + \ldots\right] \tag{141}$$

When θ is removed from 0 and π, equation (140) shows that ϕ^* becomes vanishingly small as Σ_a approaches Σ_t. On the other hand, when θ has the value 0 or π, equation (141) shows that ϕ must become very large, but it will become so at only one of these values if, at the other, ϕ and J_z have been adjusted to reduce the square-bracketed term in equation (140) to zero.

In an infinite medium of given nuclear properties, the way in which the angular flux varies with θ depends very largely on the ratio of the magnitude of the nett current density to the flux, as equation (138) shows. If the angular flux is to remain positive at all values of θ, as it clearly must do, then this ratio will be subject to a condition that is evident from equation (138) and which may be written

$$|J_z|/\phi \leq \Sigma_a \Sigma_t L^2 \frac{1 + 3(\Sigma_a/\Sigma_t)\overline{\cos\psi}\cos^2\theta}{1 + 3\Sigma_a\Sigma_t L^2 \overline{\cos\psi}} |\sec\theta|$$

The right-hand side of this condition attains its minimum value when

$$\begin{aligned} |\cos\theta| &= 1/\sqrt{[3(\Sigma_a/\Sigma_t)\overline{\cos\psi}]} \quad \text{if } (\Sigma_a/\Sigma_t)\overline{\cos\psi} > \tfrac{1}{3} \\ &= 1 \quad \text{if } (\Sigma_a/\Sigma_t)\overline{\cos\psi} \leq \tfrac{1}{3} \end{aligned}$$

so that the condition may be written

$$\left.\begin{aligned} |J_z|/\phi &\leq \Sigma_a\Sigma_t L^2 \frac{1 + 3(\Sigma_a/\Sigma_t)\overline{\cos\psi}}{1 + 3\Sigma_a\Sigma_t L^2\overline{\cos\psi}} \quad \text{if } (\Sigma_a/\Sigma_t)\overline{\cos\psi} \leq \tfrac{1}{3} \\ &\leq \Sigma_a\Sigma_t L^2 \frac{2\sqrt{[3(\Sigma_a/\Sigma_t)\overline{\cos\psi}]}}{1 + 3\Sigma_a\Sigma_t L^2\overline{\cos\psi}} \quad \text{if } (\Sigma_a/\Sigma_t)\overline{\cos\psi} > \tfrac{1}{3} \end{aligned}\right\} \quad (142)$$

The fiction of an angular flux that is negative over part of the range of θ may nevertheless be consistent with non-negative current densities J_{+z}, J_{-z}. The condition that these current densities be always positive will thus impose a less serious limitation on the magnitude of the ratio $|J_z|/\phi$ than that imposed above. From equations (130) and (132), this condition may be written

$$|J_z|/\phi \leq \tfrac{1}{2}c_z \quad (143)$$

giving an upper limit to $|J_z|/\phi$ that varies from $\frac{1}{2}$ when Σ_a/Σ_t is zero to 1 when Σ_a/Σ_t is unity.

A few distributions of angular flux are given in Figure 15.4. The full-line asymmetrical curves are those for which $|J_z|/\phi$ is only just small enough to satisfy the inequality (142). Likewise, the dotted curves are those for which $|J_z|/\phi$ is only just small enough to satisfy the inequality (143). The tendency with increased absorption for the angular flux to become small when θ is removed from 0 and π and large as θ approaches 0 or π, or both values, is clearly shown in Figure 15.4. This tendency is due to the increased absorption causing a greater contribution to the flux by neutrons that have travelled the shortest distance to the point considered. Such neutrons will be mainly those that have undergone mainly forward scattering or even no collision at all and are directed at angles θ close to 0 or π. As $\overline{\cos\psi}$

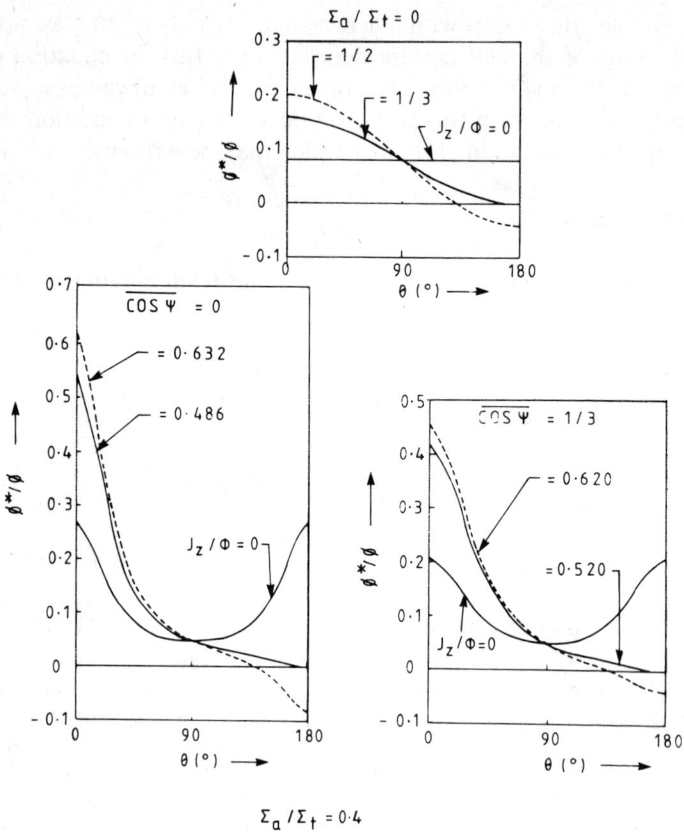

Figure 15.4 Distributions of angular flux

increases scattering collisions cause less deviation in direction so that neutrons travelling in any direction will penetrate further before being absorbed. A relatively greater contribution to the flux will then be made, as Figure 15.4 shows, by neutrons directed at angles θ further removed from 0 or π.

As concluded in Section 14.4, the current densities derived in this Chapter for any value of Σ_a/Σ_t could have been obtained by diffusion theory had scattering been isotropic. The same conclusion should also apply to the distribution of angular flux and that it does so may be shown in the following way. The probability that a neutron will be scattered from the volume element dV_1 in the direction of unit area, normal to its direction of travel, at the point (x, y, z) is given by expression (51) with unity substituted for $\mathrm{d}A \cos\theta$. Introducing this substitution and expression (54) into (53), the rate at which neutrons are scattered from volume $\mathrm{d}V_1$ so as to reach the unit area without further collision is

$$(1/4\pi)\ \phi_1\ \Sigma_s \exp(-\Sigma_t r) \sin\theta\ \mathrm{d}\varphi\ \mathrm{d}\theta\ \mathrm{d}r$$

The direction of these neutrons will lie within solid angle $\sin\theta\ \mathrm{d}\varphi\ \mathrm{d}\theta$ so that the contribution of volume $\mathrm{d}V_1$ to the angular flux $\phi^*[x, y, z, \pi - \theta, -(\pi - \varphi)]$ is

$$(1/4\pi)\ \phi_1\ \Sigma_s \exp(-\Sigma_t r)\ \mathrm{d}r$$

On replacing ϕ_1 by expression (56) and introducing equations (57), the angular flux $\phi^*(x, y, z, \theta, \varphi)$ becomes the sum of expressions such as

$$\frac{(\Sigma_s/\Sigma_t)/4\pi}{a!b!c!} \quad \frac{\partial^{a+b+c}\phi}{\partial x^a\ \partial y^b\ \partial z^c} \quad \sin^{a+b}\theta\ (-\cos\theta)^c(-\sin\varphi)^a(-\cos\varphi)^b$$

$$\times \int_0^\infty \Sigma_t\, r^{a+b+c} \exp(-\Sigma_t r)\ \mathrm{d}r$$

so that, with (59),

$$\begin{aligned}\phi^*(x, y, z, \theta, \varphi) = \frac{\Sigma_s/\Sigma_t}{4\pi} [\phi &- (1/\Sigma_t)\ \{(\partial\phi/\partial x) \sin\theta \sin\varphi \\ &+ (\partial\phi/\partial y) \sin\theta \cos\varphi + (\partial\phi/\partial z) \cos\theta\} \\ &+ (2!/\Sigma_t^2)\ \{\partial^2\phi/\partial x\ \partial y) \sin^2\theta \sin\varphi \cos\varphi \\ &+ (\partial^2\phi/\partial y\ \partial z) \sin\theta \cos\theta \cos\varphi \\ &+ (\partial^2\phi/\partial x\ \partial z) \sin\theta \cos\theta \sin\varphi\} \\ &+ (1/\Sigma_t^2)\ \{(\partial^2\phi/\partial x^2) \sin^2\theta \sin^2\varphi \\ &+ (\partial^2\phi/\partial y^2) \sin^2\theta \cos^2\varphi \\ &+ (\partial^2\phi/\partial z^2) \cos^2\theta\} - \ldots.]\end{aligned} \qquad (144)$$

If ϕ is a function of z only, then this equation becomes, with (68),

$$\begin{aligned}\phi^*(z, \theta) &= \frac{\Sigma_s/\Sigma_t}{4\pi} [\{1 + (\cos\theta/\Sigma_t L)^2 + (\cos\theta/\Sigma_t L)^4 + \\ &\quad \ldots\}\{\phi - (\cos\theta/\Sigma_t)(\mathrm{d}\phi/\mathrm{d}z)\}] \\ &= \frac{(\Sigma_s/\Sigma_t)/4\pi}{1 - (\cos\theta/\Sigma_t L)^2} [\phi - (\cos\theta/\Sigma_t)(\mathrm{d}\phi/\mathrm{d}z)] \\ &= \frac{(1/4\pi)\ \Sigma_s/\Sigma_a}{(\Sigma_t L)^2 - \cos^2\theta} [\phi(\Sigma_a\Sigma_t L^2) + J_z \cos\theta]\end{aligned} \qquad (145)$$

from (71) and (72). As expected, this result is seen to be identical to equation (138) with $\overline{\cos \psi}$ set equal to zero. A similar determination of angular flux was published by Fermi in 1936. In this case, however, derivatives higher than the first in equation (144) were ignored so that the result obtained lacked the $\cos^2 \theta$ term in the denominator, being thus restricted to a medium of sufficiently low absorption.

16

Neutron diffusion at the boundaries of a medium

16.1 Boundary conditions

The diffusion equation (68) formulates the requirement imposed on the flux ϕ by the nuclear properties of the medium through which neutrons are diffusing in order that the nett leakage of neutrons at every point shall be in balance with absorption. From physical considerations alone, however, this requirement is insufficient to determine a unique flux distribution, which must clearly depend not only on the nuclear properties of the medium but also for instance on the disposition of neutron sources and sinks at its boundaries. The flux distribution is thus subject to additional requirements known as 'boundary conditions' and an attempt to meet them may be made by adjusting certain arbitrary features that will be found to appear in the general solution to the diffusion equation. In the one-dimensional form of the diffusion equation these arbitrary features appear as a pair of arbitrary constants as, for instance, in the case of plane geometry, the constants A_1, A_2 of equation (119).

16.2 Interface between two media

A boundary condition frequently met with covers the case in which the medium A being considered is bounded by a different medium B. Referring to Figure 16.1, the interface between the two media is represented by the line $C-C$. The boundary condition relating to a point O on the interface is expressed in terms of a coordinate n, directed normally to the interface at this point from medium A to medium B. For neutrons within a particular energy range, the angular flux must be continuous across the interface for all directions Ω of neutron

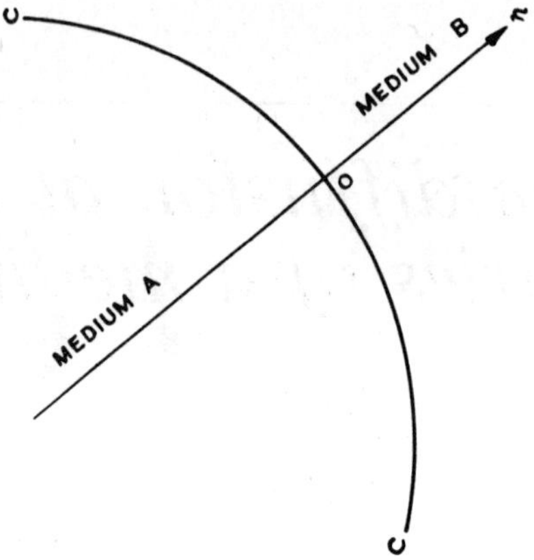

Figure 16.1 Interface between two media

travel. If, in the neighbourhood of the point O, quantities within media A and B are denoted by suffixes A and B respectively, then the required boundary condition is simply

$$\phi^*_{\mathrm{A}}(n, \Omega) = \phi^*_{\mathrm{B}}(n, \Omega) \tag{146}$$

An attempt will now be made to see how closely this condition can be satisfied by the infinite medium solution for the angular flux.

For the one-dimensional case in plane geometry the coordinate n may be identified with z. When absorption in both media is so small that terms of order $(\Sigma_a/\Sigma_t)^2$ or less may be ignored, the infinite medium solution for the angular flux ϕ^* is given by equation (139). It is clear that this solution for ϕ^* cannot be made continuous across the interface for all values of θ unless Σ_a/Σ_t is vanishingly small or is the same for each medium. Otherwise, the best that can be done is to set

$$\left.\begin{aligned} \phi_{\mathrm{A}} &= \phi_{\mathrm{B}} \\ J_{z\mathrm{A}} &= J_{z\mathrm{B}} \end{aligned}\right\} \tag{147}$$

thereby ensuring continuity of the constant and $\cos\theta$ components of ϕ^* which are independent of the absorption properties of the two media. Errors proportional to the difference in Σ_a/Σ_t for the two media must then be accepted in the case of the $(3\cos^2\theta - 1)$ and $[\cos^3\theta - (3/5)\cos\theta]$ components. Whatever the magnitude of Σ_a/Σ_t, however, it is clear from equations (102), (122) and (124) that the condition (147) will always be necessary, whether sufficient or not.

Since the infinite medium solution for ϕ or J_z in either medium contains two arbitrary constants, four are available from the two media combined. Two of these may be adjusted to meet the boundary condition across the interface, leaving the remaining two available for satisfying the boundary condition at the other face of each medium.

The inability of equation (139) to meet the interface boundary condition arises from the limitation of this equation to an infinite medium and the attempt to apply it near a boundary. However, neutron transport theory is not necessarily restricted to an infinite medium and the case of a finite medium may be dealt with by, for instance, the 'spherical harmonics' method of transport theory. It was this method that was used to obtain the relation for diffusion length involving $\overline{\cos^2 \psi}$ at the end of Section 15.4. It requires that $\phi^*(z, \theta)$ be set equal to a series of functions of θ, called spherical harmonics, having coefficients that are functions of z. The greater the number of spherical harmonics taken the higher the accuracy of the resulting solution for $\phi^*(z, \theta)$. The first four spherical harmonics are 1, $\cos \theta$, $\frac{1}{2}(3 \cos^2 \theta - 1)$ and $(5/2)\,[\cos^3 \theta - (3/5) \cos \theta]$, which may in fact be recognised in equation (139). In the resulting expression for $\phi^*(z, \theta)$, however, the coefficients, unlike those in (139), each contain a number of arbitrary constants equal to that of the spherical harmonics taken, although, for reasons that will not be gone into here, this equality is only strictly true when the number of harmonics is even. The interface boundary condition is now met by ensuring that the coefficients of corresponding harmonics are the same on either side of the interface. From the two media combined, twice as many arbitrary constants are available as are necessary for this purpose, so that half of them may be applied to the boundary condition at the other face of each medium. The infinite medium solution thus differs from that for a finite medium in that all but two of the coefficients of the harmonics are proportional to those two, the constants of proportionality depending only on the nuclear properties of the medium.

Turning to the case in which Σ_a approaches Σ_t in both media, the angular flux is given by equation (140), provided that θ is removed from 0 and π. If $\overline{\cos \psi}$ is taken to be nearly the same for each medium and if continuity of flux and nett current density is preserved across the interface, then, for all such values of θ,

$$\phi^*_A/\phi^*_B = (\Sigma_s/\Sigma_t)_A/(\Sigma_s/\Sigma_t)_B \tag{148}$$

Also, as θ approaches 0 or π, then from equation (141)

$$\phi_A^*/\phi_B^* = \frac{(\Sigma_s/\Sigma_t)_A + 2/(1 + 3\overline{\cos\psi}) + (4/2!)\,(\Sigma_t/\Sigma_s)_A/(1 + 3\overline{\cos\psi})^2 + \ldots}{(\Sigma_s/\Sigma_t)_B + 2/(1 + 3\overline{\cos\psi}) + (4/2!)\,(\Sigma_t/\Sigma_s)_B/(1 + 3\overline{\cos\psi})^2 + \ldots} \tag{149}$$

It follows that if the values of Σ_a/Σ_t in both media are near unity, a very small difference in these values will produce very large discontinuities in angular flux at the interface.

16.3 Albedo

The boundary conditions (147) can only be applied when the flux distribution is to be determined in both medium A and medium B. Adjacent media of interest, however, are necessarily limited in number so that at some of their boundaries the flux distribution is required only on one side and not on the other. A boundary such as this may conveniently be subjected to a restriction on the 'albedo' rather than to conditions (147). The albedo is a concept first introduced by Fermi and expresses the reflective property of the medium outside the boundary. It is defined as

$$\beta = \frac{\text{current density of neutrons reflected back to inside of boundary}}{\text{current density of neutrons passing to outside of boundary}} \tag{150}$$

If infinite medium diffusion theory is used to determine the albedo, then, for sufficiently small absorption, equations (74) give

$$\beta = \frac{(\phi/J_n)_b - 2}{(\phi/J_n)_b + 2} \tag{151}$$

Once β is specified for the medium outside the boundary, therefore, this equation prescribes the corresponding boundary condition. Alternatively, application of infinite medium transport theory for the one-dimensional case in plane geometry gives

$$\beta = \frac{c_z(\phi/J_z)_b - 2}{c_z(\phi/J_z)_b + 2} \tag{152}$$

from equations (130) and (132).

More generally, equations (151) and (152) may be written

$$\beta = \frac{c_n(\phi/J_n)_b - 2}{c_n(\phi/J_n)_b + 2} \tag{153}$$

where, for small absorption in three dimensions, c_n is set equal to 1 and, for the one-dimensional case in plane geometry, n is replaced by z. Also, the ratio $(\phi/J_n)_b$ is the same on either side of the boundary by conditions (147) but the coefficients c_z will be different unless the ratios of absorption to total cross-section on either side of the boundary approach one another, the mean cosine of the scattering angle being relatively unimportant. When the reflecting medium is a slab of infinite face area, the albedo may be calculated as described in Section 17.2. Results obtained in this way for various materials are listed in Appendix 6.

16.4 Medium bounded by empty space

Quite often the medium outside the boundary will be the surrounding atmosphere which, in so far as its capacity to reflect neutrons back to the inside of the boundary is concerned, may generally be regarded as a vacuum of infinite extent. In this case, the appropriate boundary condition for neutrons of one energy group travelling in any direction $\Omega(\theta, \varphi)$ is

$$\left.\begin{aligned} \pi/2 \leq \theta \leq \pi \\ \phi^*(n, \Omega)_b = 0 \end{aligned}\right\} \tag{154}$$

For the medium inside the boundary,

$$(J_{-n})_b = 0 \tag{155}$$

This relation expresses one consequence of the boundary condition (154) and so provides an approximation to that condition in terms of the infinite medium solution for ϕ. Equation (155) with (74) and (132) gives the current density of neutrons passing to the outside of the boundary as

$$(J_n)_b = -D(\partial\phi/\partial n)_b = \tfrac{1}{2}c_n\phi_b \tag{156}$$

Since the flux ϕ_b must be positive, the flux gradient $(\partial\phi/\partial n)_b$ will be negative. If, therefore, the neutron flux is extrapolated into the empty space outside the boundary to vary linearly with coordinate n, as shown in Figure 16.2, it will vanish at some distance e beyond the boundary such that

$$-e(\partial\phi/\partial n)_b = \phi_b \tag{157}$$

It then follows from equation (156) that

$$e = (2/c_n)\,D \tag{158}$$

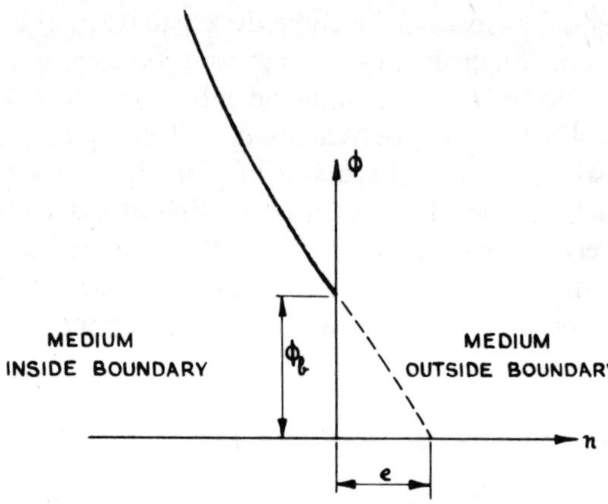

Figure 16.2 Extrapolation length

The distance e is called the 'extrapolation length'.

The surface lying outside the actual boundary surface and separated from it by the distance over which the infinite medium solution for the flux falls to zero is termed the 'extrapolated boundary'. An even simpler alternative to the boundary condition (157) is thus the condition that the flux vanishes at the extrapolated boundary. The normal distance d_b between the actual and extrapolated boundaries approximates to the extrapolation length provided that the infinite medium solution gives a negligible change in flux gradient between the two boundaries. A criterion for the validity of this approximation is readily derived for the one-dimensional case in plane geometry. In this event, the flux at the extrapolated boundary may be written

$$\phi_b + (d\phi/dz)_b d_b + (1/2!)(d^2\phi/dz^2)_b d_b^2 + (1/3!)(d^3\phi/dz^3)_b d_b^3 + \ldots = 0$$

which, with equations (68) and (157), becomes

$$\phi_b[1 - (1/e)d_b + (1/2!)(1/L^2)d_b^2 - (1/3!)(1/eL^2)d_b^3 + \ldots] = 0$$

or

$$1 + (1/2!)(d_b/L)^2 + (1/4!)(d_b/L)^4 + \ldots = (L/e)[(d_b/L) + (1/3!)(d_b/L)^3 + (1/5!)(d_b/L)^5 + \ldots]$$

so that

$$\cosh(d_b/L) = (L/e)\sinh(d_b/L)$$

and

$$d_b/L = \tanh^{-1}(e/L) \tag{159}$$

or

$$d_b = e[1 + \tfrac{1}{3}(e/L)^2 + \tfrac{1}{5}(e/L)^4 + \ldots]$$

The normal distance between the actual and extrapolated boundaries thus approaches the extrapolation length when

$$\tfrac{1}{3}(e/L)^2 \ll 1 \tag{160}$$

and this inequality is the criterion sought for. If the criterion is not satisfied, then the extrapolation length may only be used to define the boundary condition as expressed by equation (157). Alternatively, equation (159) may be used to define the extrapolated boundary more precisely.

A result similar to (158) was published by Fermi in 1936 on the basis of one-dimensional scattering. Introducing equation (83) into (82),

$$J_{-z} = \tfrac{1}{2}\overline{v}'n + \tfrac{1}{2}(1/\Sigma_t')\,\overline{v}'\ \mathrm{d}n/\mathrm{d}z$$

which, with (155) and (157), gives

$$e = 1/\Sigma_t', \tag{161}$$

agreeing with (158) for the one-dimensional case in plane geometry with large absorption on making the substitution (85). For three-dimensional scattering with small absorption, Fermi used the substitution (86) to obtain

$$e\Sigma_t = 1/\sqrt{3} \tag{162}$$

as compared with the value 2/3 given by equation (158) when $\overline{\cos\psi}$ is zero.

For the one-dimensional case in plane geometry, it follows from Section 15.6 that the condition (156) leads to negative values of the angular flux as shown by the dotted curves in Figure 15.4. Such negative values may be avoided if for the boundary condition (154) is substituted

$$\left.\begin{array}{l}\pi/2 \le \theta \le \pi\\ \text{minimum value of } \phi^*(z,\theta)_b = 0\end{array}\right\} \tag{163}$$

This condition was suggested by A. H. Wilson in 1944 and makes the current density $(J_{-n})_b$ as small as possible subject to the angular flux being always positive. From equation (138) condition (163) can only be met if $(J_z)_b$ is positive so that the alternative expressions (142) may be set equal to $(J_z)_b/\phi_b$, that is, to $1/(e/D)$. The resulting extrapolation length is an alternative to that given by equation (158) for the one-dimensional case in plane geometry and leads to angular flux distributions as given by the asymmetrical full-line curves in Figure 15.4.

Under conditions of isotropic scattering and negligible absorption, equation (158) reduces to

$$e = 2D$$

whilst equating the first of expressions (142) to $1/(e/D)$,

$$e = 3D$$

Under the same conditions, one-dimensional transport theory in plane geometry gives a flux that rises rapidly with distance inwards from the boundary and is asymptotic to an infinite medium solution. If this infinite medium solution be extended to the boundary, it will define a flux and flux gradient there from which an extrapolation length may be derived more accurately. This extrapolation length may thus be expected to lie between $2D$ and $3D$. In fact, as pointed out by Peierls in 1939, the required value of e is $2.13D$. With increasing absorption, equation (158) indicates that e/D falls, becoming unity when Σ_a is equal to Σ_t. This result accords with the more accurate derivation from transport theory for the case of isotropic scattering which was given by G. Placzek in 1949, and is quoted in the first two columns of Table 16.1. The third and fourth columns are derived from the first two with the aid of Figure 14.5. It is clear that the criterion (160) is only fulfilled when Σ_a/Σ_t is less than 0.1. A good approximation to the tabulated values of extrapolation length is seen to be

$$e/D = (2.13/c_z)[1 + 0.34(\Sigma_a/\Sigma_t) - 0.40(\Sigma_a/\Sigma_t)^2] \tag{164}$$

Expressed in this way, the extrapolation length will vary with $\cos\psi$ exactly in accordance with equation (158).

16.5 Experimental determination of extrapolation length

An experimental determination of extrapolation length, as applied to a medium bounded by empty space, was described by Amaldi and Fermi in 1936. Their experiment was made with a paraffin cylinder 240 mm in diameter and 100 mm in height. A radon–beryllium neutron source was located in this cylinder at a depth of 32 mm below the centre of the upper face. Over this depth the cylinder was cut into

Table 16.1

Σ_a/Σ_t	$e\Sigma_t$	e/D	$(e/L)^2$
0	0.7104	2.13	0.00
0.1	0.7472	2.07	0.15
0.2	0.7871	1.99	0.31
0.3	0.8300	1.90	0.47
0.4	0.8750	1.80	0.63
0.5	0.9201	1.69	0.78
0.6	0.9606	1.56	0.90
0.7	0.9889	1.41	0.98
0.8	0.9993	1.25	1.00
0.9	1.0000	1.11	1.00
1.0	1.0000	1.00	1.00

sheets normal to the axis. The neutron detector took the form of an electrolytic deposit of rhodium on a nickel plate 29 cm^2 in area, the equivalent thickness of rhodium being 0.0030 g/cm^2. It was placed between two successive paraffin sheets and irradiated until its activity had saturated. This activity was then measured by means of an ionisation chamber filled with carbon dioxide at three atmospheres pressure and connected to an electrometer. Measurements were made for various positions of the detector within the paraffin cylinder and were repeated with the detector screened on both sides by cadmium of thickness equivalent to 0.44 g/cm^2. On subtracting the activity obtained with the cadmium from that at the same position within the paraffin cylinder without the cadmium, the activity at that position due to thermal neutrons only was obtained, being directly proportional to the corresponding thermal neutron flux. The activity due to thermal neutrons was then plotted against depth of detector and an estimate made of the distance above the upper surface of the paraffin cylinder at which the extrapolated activity became zero. This distance then gave directly the extrapolation length.

17

Application of neutron diffusion theory

17.1 The diffusion theory approximation

It will perhaps appear from the last three chapters that the task of determining the flux and current density of neutrons throughout a finite medium in which they are both absorbed and anisotropically scattered is far from straightforward. Fortunately, however, a greater number of cases of interest may be dealt with to a sufficient degree of approximation by what is generally termed 'neutron diffusion theory'. This theory rests on the following statements.

(i) The relation (71) between nett current density and flux, and therewith the diffusion equation (68) can be applied throughout the medium.
(This statement is clearly in error for points in the medium within a few total mean free paths of its boundary.)

(ii) The diffusion length is given by equation (115).
(This statement has only been verified in the three-dimensional case when scattering is isotropic. With anisotropic scattering it has only been verified for the one-dimensional case in plane geometry.)

(iii) Current densities are given by equations (74) for the three-dimensional case when absorption is negligible and by (130), (132), (134) and (136) for the one-dimensional case in plane geometry whatever the absorption.
(This statement suffers from the same limitations near boundaries as statement (i). No formulae have been derived for the three-dimensional case in which absorption is significant, although equations (74) will give a good approximation provided that the absorption cross-section does not exceed about

10 per cent of the total. As is shown later, when the medium is the core of a nuclear reactor, the one-dimensional case in plane geometry gives current densities that are less dependent on cross-sections so that, by implication, equations (74) should be more generally valid in such a medium. However current densities, as opposed to nett current densities, are not normally required except at boundaries and not necessarily even then.)

(iv) At an interface between two media the flux in either medium is the same, as is also the nett current density.
(This condition is necessary but not sufficient. In the one-dimensional case in plane geometry, it has been shown to become increasingly less sufficient as the ratios of absorption to total cross-section in the two media diverge from one another.)

(v) At a boundary between a medium and empty space the extrapolation length is given by equation (164). Provided that criterion (160) is satisfied, it closely represents the normal distance between the actual and extrapolated boundary at which the flux vanishes. Otherwise, it connects the flux and flux gradient at the boundary through the relation (157). (Although this statement is true in regard to extrapolation length only for the one-dimensional case in plane geometry with isotropic scattering, it will be taken to apply generally.)

17.2 Uniformly thick slab of infinite face area

By way of example, diffusion theory will now be applied to the case in which neutrons diffuse through a slab of material that is finite and uniform in thickness but of infinite extent in planes normal thereto. Since the flux will vary only in the direction of the thickness, the theory reduces to its one-dimensional form in plane geometry.

The neutron source will be regarded as distributed over an infinite plane outside the slab and parallel to its faces. The strength of the source will be such that neutrons are emitted at a rate C per unit area of the plane and from either side of it. The neutron flux will be expressed in terms of a single coordinate z, directed along the thickness of the slab, positive in the sense from source to slab and having its origin in the face of the slab nearer the source. At its other face the slab will be deemed to be bounded by a vacuum of infinite extent.

In these circumstances the diffusion equation (68) reduces to

$$\frac{d^2\phi}{dz^2} - \frac{\phi}{L^2} = 0 \tag{165}$$

This is to be solved subject to the boundary conditions

$$\left.\begin{array}{l} J_{+z} = C \\ \text{when} \\ z = 0 \end{array}\right\} \tag{166}$$

and

$$\left.\begin{array}{l} \phi = 0 \\ \text{when} \\ z = T \end{array}\right\}$$

where T is the extrapolated thickness of the slab. It is readily verified that the required solution is

$$\phi/C = 2(L/D)\sinh\frac{T - z}{L}\left(\frac{L/D}{2/c_z} + \coth\frac{T}{L}\right)\sinh\frac{T}{L} \tag{167}$$

The ratio ϕ/C is plotted against z/L in Figure 17.1 for various values of T/L and for values of D/L equal to 0 and 0.1. At the lower value of D/L, Σ_a/Σ_t is zero so that c_z is unity. At the higher value, Σ_a/Σ_t is equal to 0.03 or less so that c_z lies between 1.000 and 1.015. In constructing Figure 17.1, c_z has been taken as unity for both values of D/L.

When T/L is infinite, equation (167) reduces to

$$\phi = \frac{4C}{c_z + 2(D/L)} e^{-z/L} \tag{168}$$

Under these circumstances no neutrons leak out by the face of the slab remote from the source and the decrease in magnitude of the flux gradient, and therewith in nett current density, as z increases is governed entirely by absorption. On the other hand, when T/L is finite, the current density into the slab at its face remote from the source will be zero. For the same flux, therefore, the nett current density at that face, and therewith the magnitude of the flux gradient, will be greater than at the corresponding value of z in a slab of infinite thickness. Continued approach to the face nearer the source is accompanied by a rapidly growing component of nett current density due to the increasing proportion of source neutrons that have so far escaped absorption, and a stage is soon reached at which this component

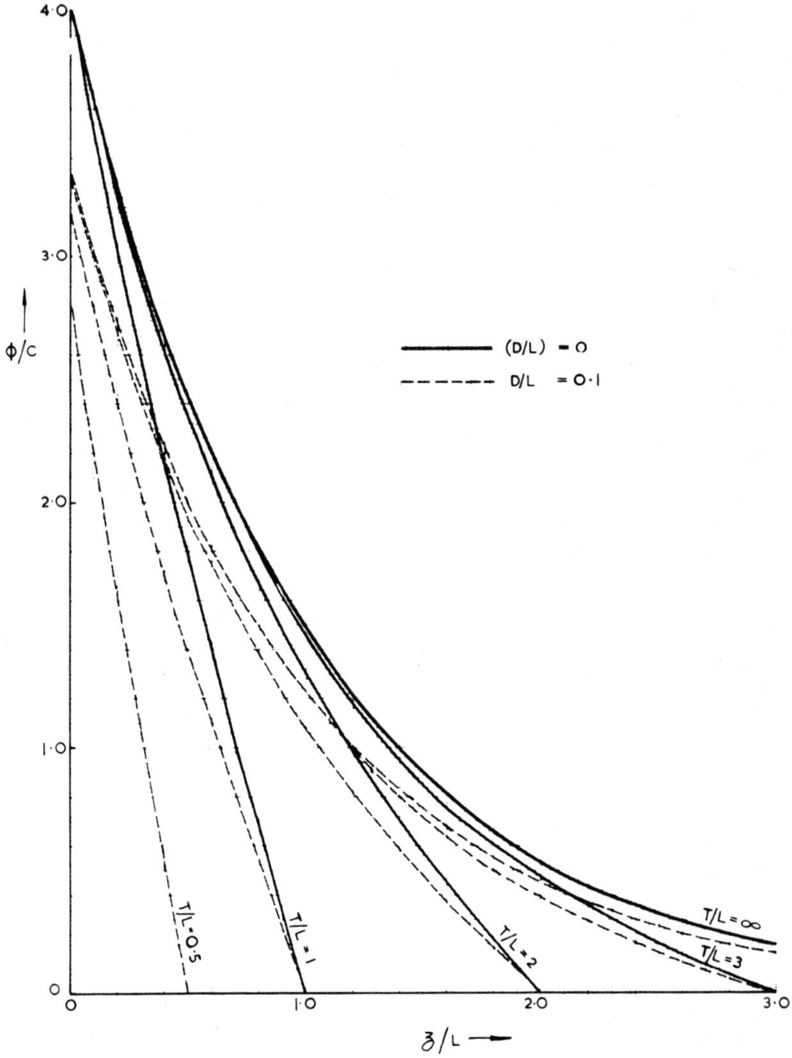

Figure 17.1 Neutron flux in slab due to plane source

swamps the unchanging leakage one. Thereafter the flux distribution becomes indistinguishable from that in a slab of infinite thickness at similar distances from the face nearer the source. At smaller values of T/L, a greater proportion of source neutrons escape by the face remote from the source so that the leakage component of the nett current density becomes dominant throughout the thickness and the flux falls almost linearly from one face to the other.

As D/L approaches zero, L becomes infinite but the case is of interest in enabling results for values of D/L between 0 and 0.1 to be obtained by interpolation. If then T remains finite, T/L will approach zero with the consequences already discussed. If, on the other hand, T/L does not approach zero, the thickness of the slab will be large. Furthermore, the flux gradient at the face remote from the source will be vanishingly small, as also will absorption throughout the slab. It follows that the nett current density at the face nearer the source will become negligible. At this face, therefore, the current density into the slab, which is C, will tend to the value $\frac{1}{4}\phi$, as in Figure 17.1, and very nearly all the neutrons entering the slab will be reflected back from it.

When D/L does not approach zero then, at the face of the slab nearer the source, the nett current density is no longer negligible. At this face, therefore, the current density into the slab is significantly greater than $\frac{1}{4}\phi$ whilst retaining its value C, so that ϕ/C is less than 4 as Figure 17.1 shows.

The source of neutrons entering the slab may be the face of an adjacent medium through which neutrons are diffusing. In this event, on applying condition (147) to the boundary between medium and slab, it follows that the ratio of flux to nett current density in the slab at its face nearer the medium, which is readily found from equation (167), may be set into (152) to give the albedo for the slab at the common boundary. This albedo is then

$$\beta = \frac{1 - (2/c_z)(D/L)\coth(T/L)}{1 + (2/c_z)(D/L)\coth(T/L)} \tag{169}$$

When T/L is infinite, the albedo attains its maximum value such that

$$\beta = \frac{1 - (2/c_z)(D/L)}{1 + (2/c_z)(D/L)} \tag{170}$$

This maximum is seen to increase as D/L diminishes, that is, as Σ_a/Σ_t diminishes.

The value of β given by equation (169) becomes close to that given by (170) when T/L is greater than 2 or so. This result is to be expected from the near coincidence of the flux distributions shown in Figure

17.1 for T/L equal to 2 and ∞ as z/L approaches zero.

In 1936, Fermi published a derivation of the albedo on the basis of one-dimensional scattering. For the case in which T/L is infinite, he obtained the result

$$\beta = \frac{1 - \sqrt{(\Sigma_a/\Sigma_t)}}{1 + \sqrt{(\Sigma_a/\Sigma_t)}} \tag{171}$$

For the corresponding case with isotropic scattering and small absorption, however, equation (170) becomes

$$\beta = \frac{1 - (2/\sqrt{3})\sqrt{(\Sigma_a/\Sigma_t)}}{1 + (2/\sqrt{3})\sqrt{(\Sigma_a/\Sigma_t)}}$$

giving a value for β smaller than that found from (171).

There remains the case in which the plane source, instead of lying outside the slab, is disposed within it. The source itself may be regarded as extending over the plane $z = 0$. The extrapolated thickness of the part of the slab for which z is positive may then be denoted by T and that for which z is negative by T'. If the source now emits neutrons at a rate C' per unit area from either side, then

$$\left.\begin{aligned} (J_{+z})_{-\delta} + C' &= (J_{+z})_{+\delta} \\ (J_{-z})_{-\delta} - C' &= (J_{-z})_{+\delta} \end{aligned}\right\}$$

where the suffix $\pm\delta$ indicates that the appropriate value of z is a small positive or negative quantity. Also if, at their common boundary, β_T is the albedo for the slab of thickness T and $\beta_{T'}$ that for the slab of thickness T', then

$$\left.\begin{aligned} (J_{+z})_{-\delta}/(J_{-z})_{-\delta} &= \beta_{T'} \\ (J_{-z})_{+\delta}/(J_{+z})_{+\delta} &= \beta_T \end{aligned}\right\}$$

It then follows that

$$(J_{+z})_{+\delta} = \frac{1 + \beta_T}{1 - \beta_T \beta_{T'}} C' \tag{172}$$

This result can be obtained in a different way by noting that for every C' neutrons that enter the slab by the face $z = \delta$ directly from the source, a further $\beta_{T'} C'$ neutrons enter by the same face after reflection from the face $z = -\delta$. Of this total of $(1 + \beta_{T'}) C'$ neutrons, a proportion $\beta_T \beta_{T'}$ will re-enter the face $z = \delta$ after reflection from both faces, and so on. Finally, therefore, it is found that

$$(J_{+z})_{+\delta} = (1 + \beta_{T'}) C' [1 + \beta_T\beta_{T'} + (\beta_T\beta_{T'})^2 + \ldots]$$

which reduces to equation (172).

From the boundary condition (166), it follows that in this case the flux at all positive values of z is given by equation (167) with C replaced by expression (172). In this expression, β_T may be replaced by the right-hand side of equation (169) and $\beta_{T'}$ replaced likewise but with T' substituted for T, so that

$$\frac{1 + \beta_{T'}}{1 - \beta_T \beta_{T'}} = \frac{[(L/D)/(2/c_z)] + \coth (T/L)}{\coth (T'/L) + \coth (T/L)} \tag{173}$$

$$\phi/C' = 2(L/D) \sinh \frac{T - z}{L} \bigg/ \left[\left(\coth \frac{T'}{L} + \coth \frac{T}{L}\right) \sinh \frac{T}{L}\right] \tag{174}$$

Comparison of this result with equation (167) shows that ϕ/C' is given by the ordinates of Figure 17.1 provided that they are first multiplied by the factor (173). If T'/L, T/L are each greater than 2 or so, and if D/L is equal to 0.1 or less, then this factor becomes approximately

$$\tfrac{1}{4}(L/D) + \tfrac{1}{2}$$

and ϕ/C' is many times greater than ϕ/C. As T' falls, however, ϕ/C' may be expected to approach ϕ/C so that the factor (173) should eventually become unity. If criterion (160) is satisfied, then this result is readily demonstrated on replacing T' by the extrapolation length as given by equation (158).

Equation (174) would naturally have been derived by direct solution of the diffusion equation (165). In this event, however, it is convenient to set down two separate solutions, one for positive and the other for negative values of z. Each solution is subject to the boundary condition that the flux vanishes at the extrapolated boundary remote from the source and is continuous across the source plane. The fourth boundary condition is that the nett current density increases by $2C'$ on passing through the source plane in the positive direction of z. The derivation of the solution for the flux in this way was published by Fermi in 1936 for the case in which T' was infinite and L^2 was written $1/(3\Sigma_t\Sigma_a)$, appropriate to small absorption with isotropic scattering.

17.3 Experimental support for diffusion theory

Experiments to verify diffusion theory were described by Amaldi and Fermi in 1936 for the case of paraffin. A radon–beryllium neutron source was embedded at a depth of 30 mm below the centre of the upper face of a paraffin cylinder. The cylinder itself measured 250 mm in diameter and 150 mm in height. Neutrons were detected by activation of a silver foil 38.5 cm^2 in area and of equivalent thickness 0.057 g/cm^2, which was placed centrally on the upper face of the paraffin cylinder. A second cylinder, identical to the first but without a neutron source, was then placed on the detector so that the two cylinders were coaxial. After measuring the activity of the detector, it was screened on both sides by cadmium of equivalent thickness 0.27 g/cm^2 and the experiment repeated. The activity of the detector in this case was subtracted from that obtained previously to give the activity A_1 due to thermal neutrons.

This experiment may be analysed in terms of an infinite medium in which thermal neutrons are generated at a uniform rate q_{th} per unit volume. The activity A_1 will then be proportional to the thermal neutron flux in such a medium so that

$$A_1 \propto q_{th}/\Sigma_a$$

In order to eliminate q_{th}, the experiment was repeated once more, but on this occasion the detector was screened by cadmium on one side only. From the resulting activity was now subtracted that obtained previously with the double-screened detector to give a second activity A_2 due to thermal neutrons.

An expression for this activity follows from a treatment similar to that published by H. A. Bethe in 1937. The thermal neutron flux in the paraffin will be reduced by the presence of the cadmium on one side of the detector. This cadmium may be likened to a negative plane source of thermal neutrons. On the assumption that all thermal neutrons entering the cadmium are absorbed, the reduction in flux in the paraffin on one side of it will be unaffected by the paraffin on the other. Equation (168) may therefore be applied on one side of the cadmium only. It follows that the reduction in flux at a distance z from the cadmium will be proportional to $e^{-z/L}$ and the flux at this distance may be written

$$\phi = (q_{th}/\Sigma_a) - Ke^{-z/L}$$

where K is some constant. To determine K, Bethe made ϕ vanish with z and then proceeded to find the current density into the cadmium. In the present instance, however, the current density

$$J_{+z} = \tfrac{1}{4}c_z\left(\frac{q_{th}}{\Sigma_a} - Ke^{-z/L}\right) - \tfrac{1}{2}(D/L)Ke^{-z/L}$$

will be made to vanish with z so that

$$K = \frac{c_z\, q_{th}/\Sigma_a}{c_z + 2(D/L)}$$

Since the activity A_2 is proportional to ϕ when z is zero, it follows that

$$A_2 \propto (q_{th}/\Sigma_a)\frac{2D/L}{c_z + 2(D/L)}$$

Finally, therefore,

$$\begin{aligned} A_1/A_2 &= \tfrac{1}{2}c_z(L/D) + 1 \qquad (175)\\ &\approx \tfrac{1}{2}c_z(L/D) \end{aligned}$$

whence, from equation (158)

$$L = e(A_1/A_2) \qquad (176)$$

In the alternative treatment adopted by Fermi and Amaldi, the probability of a neutron traversing the detector being absorbed by it was written ζ. Neutrons contributing to activity A_1 enter the detector from both sides whereas those contributing to activity A_2 can only enter the detector from one side. If, therefore, neutrons make one traverse of the detector only, the activity A_1 will be twice A_2. Of the neutrons contributing to activity A_1, however, $(1 - \zeta)/\zeta$ times as many will traverse the detector as are absorbed within it and, of these, $\beta\zeta$ will make a further contribution to activity A_1, where β is the albedo of the paraffin. Proceeding in this way, Fermi obtained the relation

$$\begin{aligned} A_1/A_2 &= 2[1 + \beta(1 - \zeta) + \beta^2(1 - \zeta)^2 + \ldots]\\ &\approx 2/(1 - \beta) \end{aligned}$$

if ζ is sufficiently small. This relation is seen to agree with the result of introducing equation (170) into (175). In order to express β in terms of nuclear cross-sections, however, equation (171) was used and lead to the result

$$\begin{aligned} A_1/A_2 &= \sqrt{(\Sigma_t/\Sigma_a)} + 1\\ &\approx \sqrt{(\Sigma_t/\Sigma_a)} \qquad (177) \end{aligned}$$

With paraffin viewed as a medium in which not only was absorption small but scattering isotropic, the diffusion length could be written

$$L = (1/\sqrt{3})\sqrt{(\Sigma_t/\Sigma_a)}/\Sigma_t$$

which, with equations (162) and (177), became identical with (176). It was thus possible to derive a value of L from measurements of the activities A_1 and A_2 and from the extrapolation length found experimentally as described in Section 16.5.

This value of L could be compared with that found from an alternative experiment performed by Amaldi and Fermi at the same time, so providing a check on the validity of the diffusion theory itself. The experiment was carried out on a paraffin cylinder 240 mm in diameter and 150 mm in height. A radon–beryllium source was embedded in this cylinder at a depth of 35 mm below the centre of its upper face. Over this depth, the cylinder was built up of laminations normal to its axis. Between any two successive laminations it was possible to insert a foil of cadmium–tin alloy, equivalent to a thickness of cadmium of 0.019 g/cm^2. This foil absorbed only thermal neutrons but was sufficiently thin to allow a proportion to pass through. The consequent reduction in flux at the upper surface of the paraffin was found by placing centrally on that surface a rhodium detector, 28 cm^2 in area and of equivalent thickness 0.36 g/cm^2, and measuring the corresponding decrease in its activity.

Since the absorbing foil itself was not activated, its neutron absorption was measured by substituting for it a rhodium plate of equal absorbing power for thermal neutrons. This equality could be checked by placing the rhodium plate between a source of slow neutrons and a detector and substracting the difference in detector activity with and without a cadmium filter from the corresponding difference without the plate. The result should be the same as the difference in activity of the unfiltered detector with and without the absorbing foil between it and the source. After inserting the rhodium plate in the paraffin, the combined activity on both sides of it was measured. The difference between this measurement and that obtained with the plate cadmium screened could then be taken as proportional to the rate of neutron uptake by the absorbing foil.

For purposes of analysis, the absorbing foil was looked upon as a negative plane source of thermal neutrons and the paraffin as a slab of finite thickness above the foil and of infinite thickness below it. On that view, the decrease A_d in activity of the rhodium detector would be proportional to the flux in such a slab at its upper face. Setting $T - z$ equal to e in equation (174), taking T' as infinite, and introducing (158),

$$\begin{aligned} A_d &\propto 2C'(L/D)\sinh[(2/c_z)(D/L)]/[1 + \coth(T/L)]\sinh(T/L) \\ &\propto (4/c_z)\, C' e^{-T/L} \end{aligned}$$

if D/L is less than about 0.1. Again, the activity A_a of the rhodium plate substituted for the absorbing foil will be proportional to the rate C' at which neutrons are emitted from unit area on either side of the negative plane source, so that

$$A_a \propto C'$$

Finally, therefore,

$$A_d/A_a \propto e^{-T/L}$$

On measuring the activities A_d and A_a for each depth of absorbing foil below the upper face of the paraffin, a plot of their ratio against this depth enabled the diffusion length L to be determined. The value so found agreed closely with that obtainable from the experiments already described, thus confirming the validity of the diffusion theory used to interpret them.

17.4 Measurement of diffusion length

A more precise measurement of diffusion length was described by H. L. Anderson and E. Fermi in 1940. In this instance the material was graphite. As shown diagrammatically in Figure 17.2, bricks of the graphite were arranged to form a block B measuring 1.22 m × 1.22 m

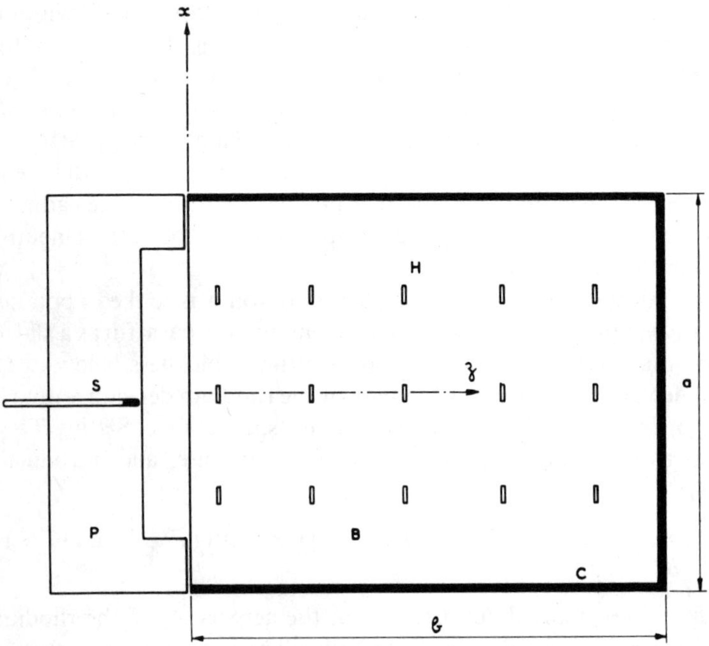

Figure 17.2 Measurement of diffusion length

× 1.525 m. The block was covered on all sides by a cadmium sheet C of effective thickness 0.45 g/cm^2. The sheet was removable on one of the square faces. Next to this face was a paraffin assembly P having at its centre a radon–beryllium neutron source S. The graphite block was traversed by 15 slots H running parallel to one of the shorter edges and provided for the introduction of a rhodium detector. The saturation activity of the detector was measured at three positions within each slot both with and without the removable cadmium sheet. At any one position, the difference in activity was proportional to the thermal neutron flux due only to thermal neutrons entering the block and not to those generated within it.

An analysis of the experiment employs a system of Cartesian coordinates x, y, z. The z-axis coincides with the axis of the block and the origin lies in the face of the block adjacent to the neutron source. The x- and y-axes run parallel to the edges of this face, the extrapolated lengths of which are a. The extrapolated length of the edge in the z-direction is b. In terms of these coordinates the diffusion equation (68) becomes

$$\frac{\partial^2\phi}{\partial x^2} + \frac{\partial^2\phi}{\partial y^2} + \frac{\partial^2\phi}{\partial z^2} - \frac{\phi}{L^2} = 0$$

and is to be solved subject to the boundary conditions

$$\phi = 0$$

when

$$x = \tfrac{1}{2}a$$
$$y = \tfrac{1}{2}a$$
$$z = b$$

It may readily be verified that a solution that satisfies these conditions is

$$\phi = C_{mn} \cos \frac{m\,\pi\,x}{a} \cos \frac{n\,\pi\,y}{a} [\exp(-\gamma_{mn}z)$$
$$- \exp(-2\gamma_{mn}b \,.\, \exp(\gamma_{mn}z)]$$

provided that

$$\gamma_{mn}^2 = (1/L^2) + (\pi/a)^2 (m^2 + n^2) \qquad (178)$$

where

$$m, n = 1, 3, 5, \ldots$$

and C_{mn} is an arbitrary constant. Likewise any sum of such solutions will itself be a solution so that, more generally,

$$\phi = \sum_{m=1}^{\infty} \sum_{n=1}^{\infty} C_{mn} \cos \frac{m\pi x}{a} \cos \frac{n\pi y}{a} \left[\exp(-\gamma_{mn} z) - \exp(-2\gamma_{mn} b) \, . \, \exp(\gamma_{mn} z) \right] \tag{179}$$

The activity proportional to the thermal neutron flux was determined at nine points in each measuring plane normal to the z-axis. A linear combination of these measurements was chosen so as to minimise the influence of the higher harmonics in equation (179) and leave the term for which m and n are unity predominant. This term was in fact found to be by far the most significant. Its variation with z was now known and could be equated to

$$\exp(-\gamma_{11} z) - \exp(-2\gamma_{11} b) \quad \exp(\gamma_{11} z)$$

to give γ_{11}. The required value of L was finally found from equation (178) which, in this case, became

$$\gamma_{11}^2 = (1/L^2) + 2(\pi/a)^2.$$

The diffusion length thus obtained corresponded to an absorption cross-section which could be accounted for by a water content of less than 1 per cent. This possibility was excluded by comparing the weights of some of the graphite bricks before and after heating at 200 °C. There was a further possibility of the gaps between the micro-crystals in the graphite being filled with air. In this event, however, the nitrogen in the air would account for only some 10 per cent of the absorption cross-section observed.

Current methods of measuring diffusion length do not differ widely from that just described. More recent measurements on a selection of materials are listed in Appendix 6.

18

Slowing down of neutrons from a source of infinite extent

18.1 Neutrons at once-collided energies

The spatial distribution of neutron flux has so far been discussed with reference to neutrons having energies close to some average value, in particular 'thermal' energies. In general, however, the neutrons emanating from a source are fast neutrons and it is only as a result of their slowing down that neutrons of thermal energies are produced. During the slowing-down process diffusion will take place as in the case of thermal neutrons. Description of the behaviour of fast neutrons thus poses the problem of expressing the flux per unit energy ϕ' as a function of energy and space. This problem was receiving attention at the same time as that of neutron diffusion, notably by E. Fermi, E. Amaldi and G. Placzek.

In the first instance, discussion will be restricted to conditions under which the flux per unit energy varies only with energy and is spatially uniform. These conditions obtain for neutrons slowing down from a source uniformly distributed throughout a medium of infinite extent. The neutrons will be supposed to originate from such a source at a rate q_0 per unit volume of the medium and at an energy E_0. The proportion of the source neutrons undergoing scattering collisions is then

$$\left(\frac{\Sigma_s}{\Sigma_s + \Sigma_a}\right)_{E_0}$$

and the lowest energy to which they can be reduced by their first such collision is αE_0, where α is given by equation (31) and may be read from Figure 18.1. However, referring to Figure 18.2, a proportion of these scattered neutrons will be degraded into a range dE about an energy E greater than αE_0. From equation (32) this proportion is

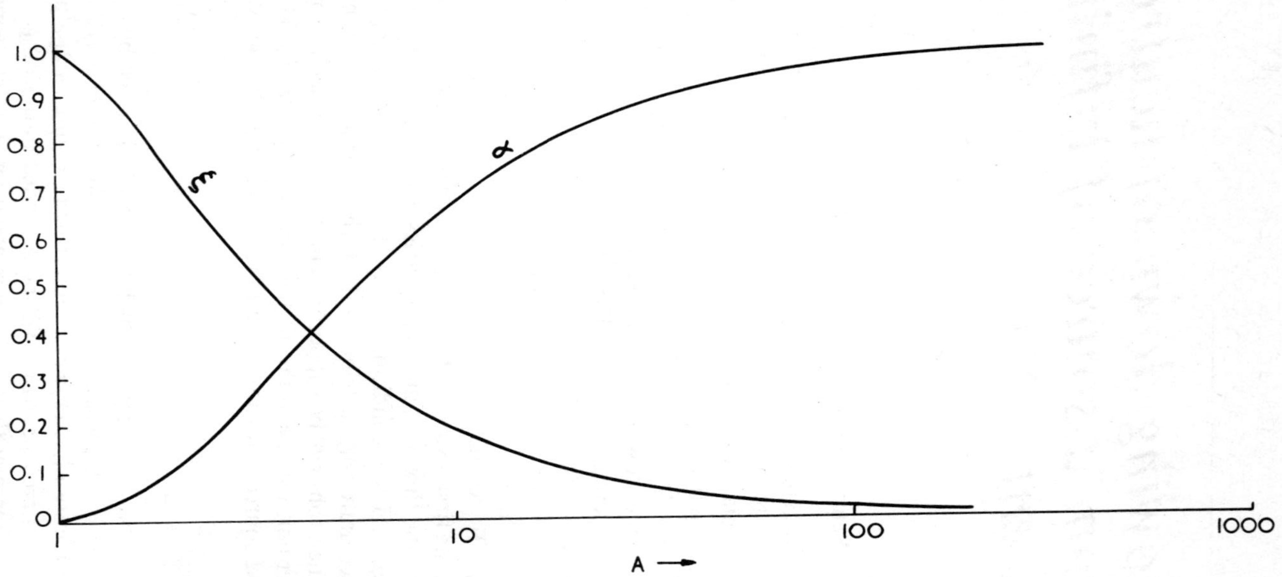

Figure 18.1 Maximum energy reduction and mean logarithmic energy loss per collision

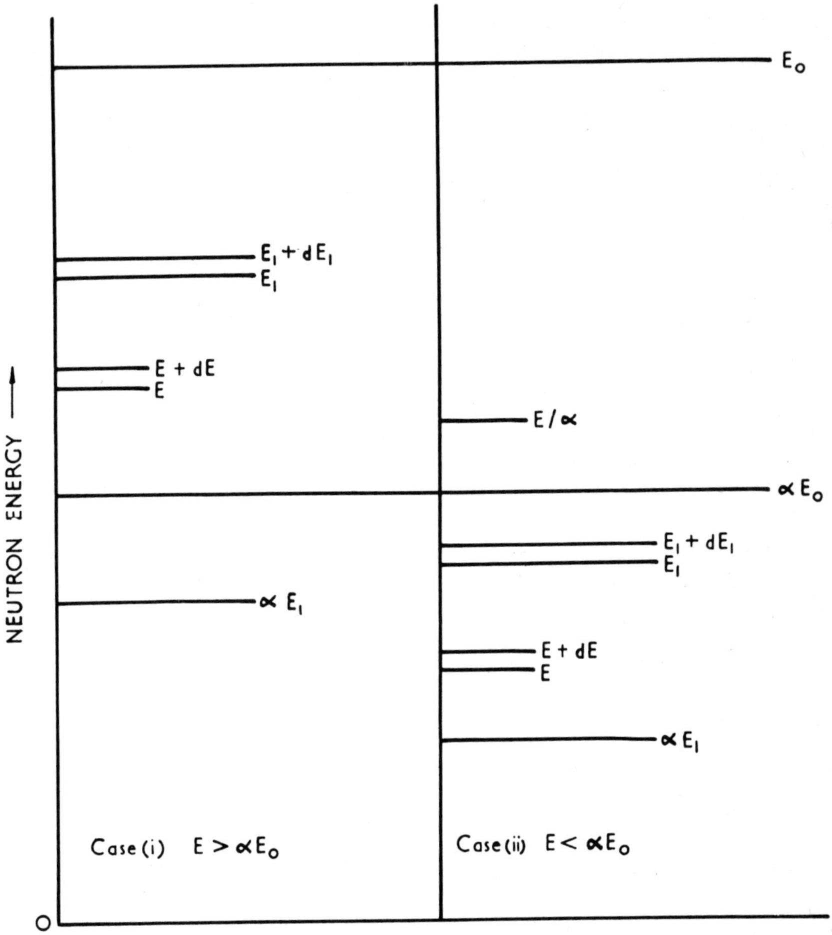

Figure 18.2 Slowing down from a source of infinite extent

$$\mathrm{d}E/(1 - \alpha)E_0$$

if scattering is isotropic in the centre-of-mass system. The rate per unit volume at which source neutrons are scattered into energy range dE is therefore

$$\left(\frac{\Sigma_s}{\Sigma_s + \Sigma_a}\right)_{E_0} q_0 \frac{\mathrm{d}E}{(1 - \alpha)E_0} \tag{180}$$

This expression is limited to neutrons scattered directly from the source energy into range dE as a result of one collision only. It is clear, however, that neutrons scattered into a range such as dE_1, about an energy E_1 higher than E, may also enter range dE. The rate per unit volume at which neutrons are scattered out of range dE_1 is

$$(\Sigma_s)_{E_1}\,\phi'(E_1)\,dE_1$$

and the proportion of such neutrons entering range dE is

$$dE/(1-\alpha)\,E_1$$

The rate per unit volume at which all neutrons making more than one collision are scattered into energy range dE is therefore

$$\frac{dE}{1-\alpha}\int_E^{E_0}(\Sigma_s)_{E_1}\,\phi'(E_1)\frac{dE_1}{E_1} \qquad (181)$$

Finally, the combined rate per unit volume at which neutrons are absorbed in and scattered out of energy range dE is

$$(\Sigma_s + \Sigma_a)_E\,\phi'(E)\,dE$$

If the density of neutrons within the range dE is not to vary with time, then this expression must be equal to the sum of (180) and (181) so that

$$(\Sigma_s + \Sigma_a)_E\,\phi'(E) = \frac{1}{1-\alpha}\left[\left(\frac{\Sigma_s}{\Sigma_s+\Sigma_a}\right)_{E_0}(q_0/E_0) + \int_E^{E_0}(\Sigma_s)_{E_1}\,\phi'(E_1)\frac{dE_1}{E_1}\right] \qquad (182)$$

where

$$E > \alpha E_0$$

To solve equation (182), it is first differentiated with respect to E to give

$$\frac{d}{dE}\left[(\Sigma_s + \Sigma_a)_E\,\phi'(E)\right]$$

$$= -\frac{1}{1-\alpha}\left(\frac{\Sigma_s}{\Sigma_s+\Sigma_a}\right)_E(\Sigma_s + \Sigma_a)_E\,\phi'(E)/E \qquad (183)$$

In this form it may be directly integrated to yield the solution

$$(\Sigma_s + \Sigma_a)_E\,\phi'(E)$$

$$= \left(\frac{\Sigma_s}{\Sigma_s+\Sigma_a}\right)_E\frac{q_0/E_0}{1-\alpha}\,(E_0/E)^{1/(1-\alpha)}\ \times$$

$$\exp\left[-\frac{1}{1-\alpha}\int_{E}^{E_0}\left(\frac{\Sigma_a}{\Sigma_s+\Sigma_a}\right)_{E_1}\frac{dE_1}{E_1}\right] \tag{184}$$

which satisfies the boundary condition imposed on $\phi'(E)$ at energy E_0 by equation (182).

When α is zero, as is very nearly the case for hydrogen, equation (184) applies over the whole energy range through which the neutrons are slowed down and may be written

$$(\Sigma_s+\Sigma_a)_E\,\phi'(E) = \left(\frac{\Sigma_s}{\Sigma_s+\Sigma_a}\right)_{E_0}\frac{q_0}{E}\exp\left[-\int_{E}^{E_0}\left(\frac{\Sigma_a}{\Sigma_s+\Sigma_a}\right)_{E_1}\frac{dE_1}{E_1}\right] \tag{185}$$

For the case of zero absorption this result was quoted by Amaldi and Fermi in 1936. With increasing values of α, the energies of once-collided neutrons become higher so that absorption is generally very small. In this event equation (184) reduces to

$$(\Sigma_s)_E\,\phi'(E) = \frac{q_0/E_0}{1-\alpha}(E_0/E)^{1/(1-\alpha)} \tag{186}$$

For various values of α and in the absence of absorption, $(\Sigma_s)_E\,\phi'(E)$ is shown plotted against E/E_0 in Figure 18.3. Only for values of E/E_0 greater than α, however, are the curves given by equation (186). Equation (182), and its solution (186) for the case of zero absorption, were derived by Placzek in 1940.

18.2 Neutrons below once-collided energies

Consideration will next be given to the slowing down of neutrons past energies lower than those attainable by one collision alone. In this case, the expression for the rate per unit volume at which neutrons are scattered into a range dE about an energy E, which is now less than αE_0, will contain no term analogous to (180). It will however contain a term identical to (181) apart from the upper limit, which is now E/α instead of E_0. The equation corresponding to (182) thus becomes

$$\left.\begin{aligned}&(\Sigma_s+\Sigma_a)_E\,\phi'(E) = \frac{1}{1-\alpha}\int_{E}^{E/\alpha}(\Sigma_s)_{E_1}\,\phi'(E_1)\frac{dE_1}{E_1}\\ &\text{where}\\ &E < \alpha E_0\end{aligned}\right\} \tag{187}$$

On setting E equal to αE_0 in both equations (182) and (187), their difference gives

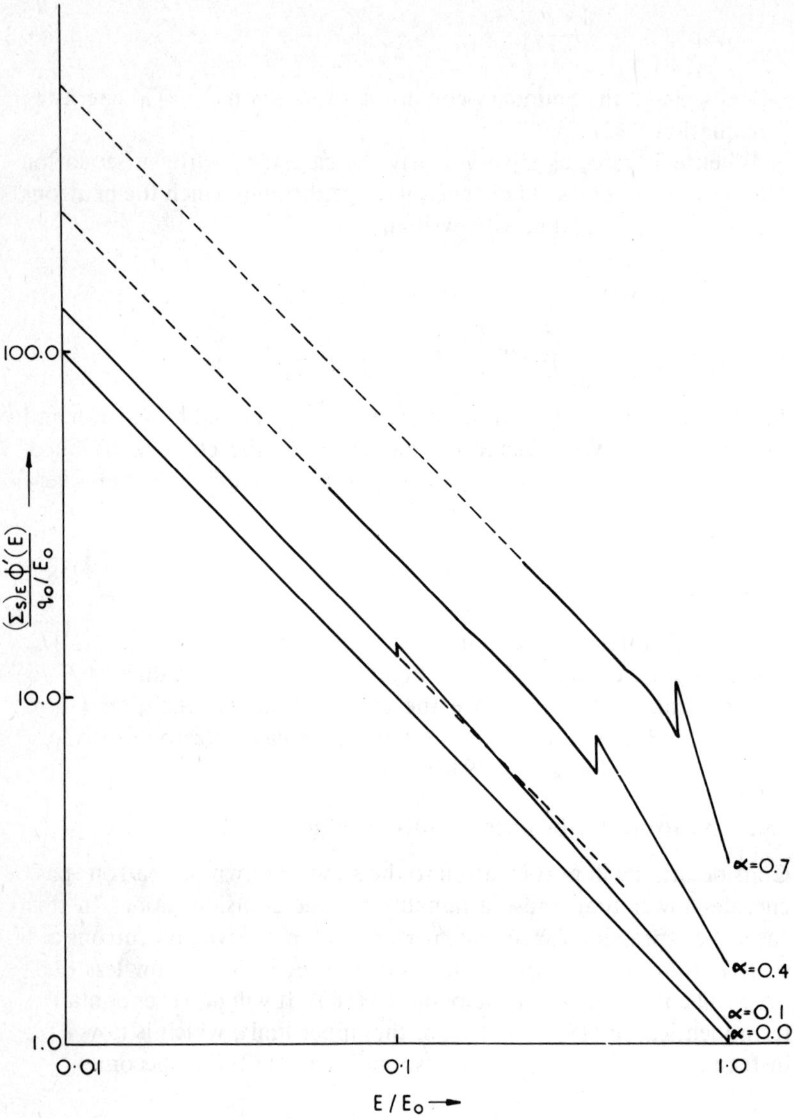

Figure 18.3 Flux during slowing down

$$(\Sigma_s + \Sigma_a)_{\alpha E_0}\,[\phi'(\alpha E_0 + dE) - \phi'(\alpha E_0 - dE)]$$

$$= \frac{1}{1-\alpha}\left(\frac{\Sigma_s}{\Sigma_s + \Sigma_a}\right)_{E_0} q_0/E_0 \qquad (188)$$

Thus, at energy αE_0, the flux per unit energy experiences a sudden fall which increases with α, as shown in Figure 18.3.

In a given small range of energy, a proportion of the flux is due to neutrons that have made more than one collision and this proportion will increase as the energy level of the range falls. The remaining proportion arises from once-collided neutrons and is unchanged in amount at energies down to αE_0 but zero thereafter. It is this sudden disappearance of the once-collided neutrons that gives rise to the discontinuity in flux at energy αE_0.

Equation (187) may be solved as described by Placzec in 1940. It is first differentiated with respect to E to give

$$\frac{d}{dE}[(\Sigma_s + \Sigma_a)_E\,\phi'(E)]$$

$$= \frac{(\Sigma_s)_{E/\alpha}\phi'(E/\alpha) - (\Sigma_s)_E\,\phi'(E)}{E(1-\alpha)} \qquad (189)$$

If absorption be discounted, then this equation has the solution

$$\left.\begin{aligned}&(\Sigma_s)_E\,\phi'(E)\\ &= (\alpha^n E_0/E)^{1/(1-\alpha)}(\Sigma_s)_{\alpha^n E_0}\phi'(\alpha^n E_0)\\ &\quad - \frac{E^{-1/(1-\alpha)}}{1-\alpha}\int_E^{\alpha^n E_0} E_1^{\,1/(1-\alpha)}\,(\Sigma_s)_{E_1/\alpha}\phi'(E_1/\alpha)\,\frac{dE_1}{E_1}\end{aligned}\right\} \qquad (190)$$

provided that

$$E < \alpha^n E_0$$

where n is an integer. In order to avoid discontinuity in the integrand E will be subjected to a lower bound $\alpha^{n+1}E_0$.

For the case in which n is 1, it follows from equation (186) that

$$(\Sigma_s)_{E_1/\alpha}\phi'(E_1/\alpha) = \frac{q_0/E_0}{1-\alpha}\,(\alpha E_0/E_1)^{1/(1-\alpha)}$$

and, from (186) and (188), that

$$(\Sigma_s)_{\alpha E_0}\,\phi'(\alpha E_0) = \frac{q_0/E_0}{1-\alpha}\left[\left(\frac{1}{\alpha}\right)^{1/(1-\alpha)} - 1\right]$$

With these last two relations equation (190) becomes

$$(\Sigma_s)_E\,\phi'(E) = \frac{q_0/E_0}{1-\alpha}(E_0/E)^{1/(1-\alpha)}\left[1 - \alpha^{1/(1-\alpha)}\left\{1+\left(\frac{\log(\alpha E_0/E)}{1-\alpha}\right)\right\}\right] \tag{191}$$

For the case in which n is 2. equation (191) gives

$$\left.\begin{aligned}(\Sigma_s)_{E_1/\alpha}\,\phi'(E_1/\alpha) &= \frac{q_0/E_0}{1-\alpha}(\alpha E_0/E_1)^{1/(1-\alpha)}\left[1 - \alpha^{1/(1-\alpha)}\left(1+\frac{\log(\alpha^2E_0/E_1)}{1-\alpha}\right)\right]\\ (\Sigma_s)_{\alpha^2E_0}\,\phi'(\alpha^2E_0) &= \frac{q_0/E_0}{1-\alpha}(1/\alpha^2)^{1(1-\alpha)}\left[1 - \alpha^{1/(1-\alpha)}\left(1+\frac{\log(1/\alpha)}{1-\alpha}\right)\right]\end{aligned}\right\}$$

whence (190) becomes

$$(\Sigma_s)_E\,\phi'(E) = \frac{q_0/E_0}{1-\alpha}(E_0/E)^{1/(1-\alpha)}\left[1 - \alpha^{1/(1-\alpha)}\left\{1+\left(\frac{\log(\alpha E_0/E)}{1-\alpha}\right)\right\} + (\alpha^2)^{1/(1-\alpha)}\left\{\left(\frac{\log(\alpha^2E_0/E)}{1-\alpha}\right)+\frac{1}{2}\left(\frac{\log(\alpha^2E_0/E)}{1-\alpha}\right)^2\right\}\right]$$

Proceeding in this way the result for any value of n is found to be

$$\begin{aligned}(\Sigma_s)_E\,\phi'(E) &= \frac{q_0/E_0}{1-\alpha}(E_0/E)^{1/(1-\alpha)}\left[1 - \alpha^{1/(1-\alpha)}\left\{1+\left(\frac{\log(\alpha E_0/E)}{1-\alpha}\right)\right\}\right.\\ &+ (\alpha^2)^{1/(1-\alpha)}\left\{\left(\frac{\log(\alpha^2E_0/E)}{1-\alpha}\right)+\frac{1}{2!}\left(\frac{\log(\alpha^2E_0/E)}{1-\alpha}\right)^2\right\}\\ &- (\alpha^3)^{1/(1-\alpha)}\left\{\frac{1}{2!}\left(\frac{\log(\alpha^3E_0/E)}{1-\alpha}\right)^2+\frac{1}{3!}\left(\frac{\log(\alpha^3E_0/E)}{1-\alpha}\right)^3\right\}\\ &+ \ldots\\ &+ (-1)^n(\alpha^n)^{1/(1-\alpha)}\left\{\frac{1}{(n-1)!}\left(\frac{\log(\alpha^nE_0/E)}{1-\alpha}\right)^{n-1}\right.\end{aligned}$$

$$+ \frac{1}{n!}\left(\frac{\log(\alpha^n E_0/E)}{1-\alpha}\right)^n\Bigg\}\Bigg] \qquad (192)$$

This equation may be used to obtain the flux per unit energy at energies below αE_0, as shown by the full lines in Figure 18.3.

In a given small range of energy the contribution to the flux from twice-collided neutrons will gradually diminish as the energy level falls below αE_0, whilst that from neutrons that have collided more frequently will continue to grow. At energy $\alpha^2 E_0$ the contribution from twice-collided neutrons will have disappeared altogether. As this disappearance occurs gradually and not suddenly it causes a discontinuity not in the flux itself but in its rate of change with energy. In the same way, the contribution from thrice-collided neutrons will have disappeared at energy $\alpha^3 E_0$. This contribution builds up and dies away over a greater energy interval than that of the twice-collided neutrons so that its disappearance is even more gradual. The consequent discontinuity in rate of change of flux with energy is correspondingly less marked, as Figure 18.3 shows.

18.3 Asymptotic slowing down

With decreasing energy, the flux per unit energy is seen from Figure 18.3 to approach an inverse variation with the product $(\Sigma_s)_E E$, as shown by the dotted lines. This result may be deduced from equation (189) rewritten in the form

$$\frac{d}{dE}\left\{E\left(\Sigma_s + \Sigma_a\right)_E \phi'(E)\right\}$$

$$= \frac{(E/\alpha)(\Sigma_s)_{E/\alpha}\phi'(E/\alpha) - E(\Sigma_s)_E\phi'(E)}{(E/\alpha) - E}$$

$$+ (\Sigma_a)_E\phi'(E) \qquad (193)$$

On introducing $dE = -E^2\, d(1/E)$

equation (193) becomes

$$\frac{d}{d(1/E)}\left[E\left(\Sigma_s + \Sigma_a\right)_E \phi'(E)\right]$$

$$= \alpha\,\frac{E(\Sigma_s)_E\phi'(E) - (E/\alpha)(\Sigma_s)_{E/\alpha}\phi'(E/\alpha)}{(1/E) - (\alpha/E)}$$

$$- (1/E^2)(\Sigma_a)_E\phi'(E)$$

With Σ_a set equal to zero then, as pointed out by Placzek in 1940, the quantity $E\,(\Sigma_s)_E\phi'(E)$ is seen to have a gradient at $1/E$ that is a fraction α of its average gradient over the range $1/E$ to α/E immediately below $1/E$. With increasing values of $1/E$, therefore, the gradient will approach zero. At a sufficiently low energy, which may be expected to be less than that given by

$$E/\alpha = \alpha E_0$$

so that it is less than $\alpha^2 E_0$, it follows that

$$E(\Sigma_s)_E\phi'(E) = h_0 \qquad (194)$$

where h_0 is a constant. Slowing down is then said to have become 'asymptotic'.

The constant h_0 in equation (194) cannot be determined by substituting (194) into (187) since it will merely cancel on both sides. Instead, equation (193) is first integrated between limits E and αE_0 to give

$$\begin{aligned}\int_E^{\alpha E_0} \mathrm{d}\left\{E_1(\Sigma_s + \Sigma_a)_{E_1}\phi'(E_1)\right\} &= \frac{\alpha}{1-\alpha}\left\{\int_{E/\alpha}^{E_0}(\Sigma_s)_{E_1}\phi'(E_1)\mathrm{d}E_1 - \int_E^{\alpha E_0}(\Sigma_s)_{E_1}\phi'(E_1)\mathrm{d}E_1\right\} \\ &\quad + \int_E^{\alpha E_0}(\Sigma_a)_{E_1}\phi'(E_1)\mathrm{d}E_1\end{aligned}$$

whence

$$\begin{aligned}\alpha E_0\,(\Sigma_s + \Sigma_a)_{\alpha E_0}\phi'\,(\alpha E_0 - \delta E) &= E(\Sigma_s + \Sigma_a)_E\phi'(E) \\ &\quad + \frac{\alpha}{1-\alpha}\left\{\int_{\alpha E_0}^{E_0}(\Sigma_s)_{E_1}\phi'(E_1)\mathrm{d}E_1 - \int_E^{E/\alpha}(\Sigma_s)_{E_1}\phi'(E_1)\mathrm{d}E_1\right\} \\ &\quad + \int_E^{\alpha E_0}(\Sigma_a)_{E_1}\phi'(E_1)\mathrm{d}E_1 \qquad (195)\end{aligned}$$

For energies greater than αE_0, however, equation (183) may be written

$$\frac{\mathrm{d}}{\mathrm{d}E}\left\{E(\Sigma_s + \Sigma_a)_E\phi'(E)\right\}$$

$$= -\frac{1}{1-\alpha}(\Sigma_s)_E\phi'(E) + (\Sigma_s + \Sigma_a)_E\phi'(E)$$

$$= -\frac{\alpha}{1-\alpha}(\Sigma_s)_E\phi'(E) + (\Sigma_a)_E\phi'(E)$$

which, on integration, becomes

$$E_0(\Sigma_s + \Sigma_a)_{E_0}\phi'(E_0) - \alpha E_0(\Sigma_s + \Sigma_a)_{\alpha E_0}\phi'(\alpha E_0 + \delta E)$$

$$= -\frac{\alpha}{1-\alpha}\int_{\alpha E_0}^{E_0}(\Sigma_s)_{E_1}\phi'(E_1)\mathrm{d}E_1 + \int_{\alpha E_0}^{E_0}(\Sigma_a)_{E_1}\phi'(E_1)\mathrm{d}E_1 \qquad (196)$$

Adding equations (195) and (196) and introducing (188),

$$E_0(\Sigma_s + \Sigma_a)_{E_0}\phi'(E_0) - \frac{\alpha}{1-\alpha}\left(\frac{\Sigma_s}{\Sigma_s + \Sigma_a}\right)_{E_0}q_0$$

$$= E(\Sigma_s + \Sigma_a)_E\phi'(E) - \frac{\alpha}{1-\alpha}\int_E^{E/\alpha}(\Sigma_s)_{E_1}\phi'(E_1)\mathrm{d}E_1 + \int_E^{E_0}(\Sigma_a)_{E_1}\phi'(E_1)\mathrm{d}E_1$$

which, from (182) with E set equal to E_0, takes the form

$$\left(\frac{\Sigma_s}{\Sigma_s + \Sigma_a}\right)_{E_0}q_0 - \int_E^{E_0}(\Sigma_a)_{E_1}\phi'(E_1)\mathrm{d}E_1$$

$$= E(\Sigma_s + \Sigma_a)_E\phi'(E) - \frac{\alpha}{1-\alpha}\int_E^{E/\alpha}(\Sigma_s)_{E_1}\phi'(E_1)\mathrm{d}E_1 \qquad (197)$$

where

$$E < \alpha E_0$$

Equation (187) and its transformation (197) were derived by Placzek in 1940.

For the case of asymptotic slowing down, equation (194) may be set into (197), with zero absorption, to give

$$h_0 = q_0/\xi$$

where

$$\xi = 1 - [\alpha/(1-\alpha)]\log(1/\alpha) \qquad (198)$$

Finally, therefore, equation (194) becomes

$$(\Sigma_s)_E\phi'(E) = q_0/\xi E \tag{199}$$

as published by Placzek in 1939.

The quantity ξ is shown plotted against mass number A in Figure 18.1. It can be physically interpreted as the 'mean logarithmic energy loss per collision'. For this quantity is, from equation (32),

$$\int_{\alpha E'}^{E'} (\log E' - \log E)\frac{\mathrm{d}E}{(1-\alpha)E'}$$

$$= \frac{1}{1-\alpha}\int_{E'}^{\alpha E'} \log(E/E')\mathrm{d}(E/E')$$

$$= 1 - [\alpha/(1-\alpha)]\log(1/\alpha)$$

which is identical to the expression given for ξ by (198).

A neutron scattered through a given angle has its energy reduced in a definite proportion and therefore loses more energy in earlier than in later collisions. It thus makes fewer collisions in falling through an energy range at a higher energy level than in falling through the same range at a lower level. It follows that the rate of scattering per unit volume and energy is less at higher energies, as equation (199) shows. Again, the greater the value of ξ, the more energy a neutron loses on collision and the fewer collisions it makes in falling through a given energy range. The smaller is then the rate of scattering per unit volume and energy, as also appears from equation (199).

If equation (199) is multiplied on both sides by the small energy range $\mathrm{d}E$, then the term

$$(\mathrm{d}E/E)/\xi$$

on the right-hand side is seen to represent the probability that a neutron slowing down through the energy range $\mathrm{d}E$ will make a collision in the range. This representation becomes clearer on rewriting the term in the form

$$\mathrm{d}(\log E)/\xi$$

Since ξ is the logarithmic energy loss of a neutron per collision, the term in this form must express the proportion of a collision made by a neutron, or the proportion of neutrons making collisions, in slowing down through the energy range $\mathrm{d}E$.

18.4 Slowing-down density

The left-hand side of equation (197) is the rate per unit volume at which neutrons are slowed down past energy E. This rate is known as the 'slowing-down density' and is denoted by q. On introducing equation (187), it follows that (197) may be written in the form

$$q = \int_{E}^{E/\alpha} (\Sigma_s)_{E_1} \phi'(E_1) \frac{E - \alpha E_1}{E_1(1 - \alpha)} \, dE_1 \tag{200}$$

The fraction

$$\frac{E - \alpha E_1}{E_1(1 - \alpha)}$$

is the probability that a neutron of energy E_1 will be scattered into the range of energy below E and, with this realisation, equation (197) might have been set up directly in the form (200).

The slowing-down density given by equation (200) does not apply at once-collided energies. At these energies the analogous equation is

$$q = \left(\frac{\Sigma_s}{\Sigma_s + \Sigma_a}\right)_{E_0} q_0 \frac{E - \alpha E_0}{E_0(1 - \alpha)} + \int_{E}^{E_0} (\Sigma_s)_{E_1} \phi'(E_1) \frac{E - \alpha E_1}{E_1(1 - \alpha)} \, dE_1 \tag{201}$$

where

$$E > \alpha E_0$$

If, in equations (200) and (201), E is set equal to αE_0, it is seen that there is no discontinuity in q. This result is to be expected, despite the accompanying discontinuity in $\phi'(E)$.

18.5 Slowing down by atoms of several types

In the case of a material formed of several different types of atom denoted by suffix i = 1,2, . . ., equation (200) may be written

$$q = \sum_{i=1,2,\ldots} \frac{\Sigma_{s_i}}{\Sigma_s} \int_{E}^{E/\alpha_i} (\Sigma_s)_{E_1} \phi'(E_1) \frac{E - \alpha_i E_1}{E_1(1 - \alpha_i)} \, dE_1 \tag{202}$$

where the cross-section Σ_{s_i} for scattering collisions with atoms of type i is taken to vary with neutron energy in the same way as the scattering cross-section Σ_s for the material as a whole. In the absence of absorption, the slowing-down density q becomes the source strength q_0. Also, for the case of asymptotic slowing down, equation (194) may be introduced into (202) so that the integration may be carried out to give

$$q_0 = h_0 \sum_{i=1,2,\ldots} (\Sigma_{s_i}/\Sigma_s) \frac{1}{1-\alpha_i} \left[-\frac{E}{E_1} - \alpha_i \log E_1 \right]_E^{E/\alpha_i}$$

$$= h_0 \sum_{i=1,2,\ldots} (\Sigma_{s_i}/\Sigma_s) \left[1 - \frac{\alpha_i}{1-\alpha_i} \log (1/\alpha_i) \right]$$

$$= h_0 \sum_{i=1,2,\ldots} (\Sigma_{s_i}/\Sigma_s)\, \xi_i$$

from (198). It follows that

$$h_0 = q_0/\bar{\xi}$$

where

$$\bar{\xi} = \sum_{i=1,2,\ldots} (\Sigma_{s_i}/\Sigma_s)\, \xi_i \tag{203}$$

and equation (194) becomes

$$(\Sigma_s)_E \phi'(E) = q_0/\bar{\xi} E$$

analogous to (199).

It should be noted that expression (203) has only been derived for the case of asymptotic slowing down and is only valid if the scattering cross-sections of all atoms in the material vary with energy in exactly the same way.

19

Resonance capture

19.1 Neutron energies for resonance capture

Early in the year 1935, P. B. Moon and J. R. Tillman observed that the absorption of slow neutrons by a block of iodine appeared to be greater from measurements of neutron transmission made with an iodine detector than from those made with detectors of silver or rhodium. A year later, similar observations had been made by Amaldi and Fermi using absorbers and detectors of rhodium and silver. With a rhodium detector the absorption of slow neutrons by rhodium was apparently greater than that by silver and with a silver detector the opposite result was obtained. Moreover, the effect was enhanced when the source of slow neutrons was filtered by cadmium. It thus appeared that a material might exhibit high absorption cross-sections for slow neutrons within particular energy ranges, higher than the thermal one. The neutrons reaching a detector of the same material would then be much less readily absorbed by it. Absorption of this kind is referred to as 'resonance capture'.

The first attempt to determine the neutron energies at which resonance capture occurred in several different elements was made by H. H. Goldsmith and F. Rasetti in 1936. Their neutron source was a bulb containing 300–600 millicuries of radon and powdered beryllium. It was placed at the centre of a paraffin block of dimensions 300 × 220 × 75 mm, which was completely covered by cadmium of effective thickness 0.4 g/cm^2. The detectors, measuring 40 × 50 mm, were placed at the centre of the largest face of the paraffin block and at a distance of about 10 mm from the cadmium, sufficient to permit the interposition of absorbers. The absorbers were Pyrex plates equivalent to 0.020 or 0.027 g/cm^2 effective thickness of boron, boron carbide equivalent to 0.58 g/cm^2 effective thickness of boron, or lithium fluoride equivalent to 0.22 g/cm^2 effective thickness of lithium. Absorption by elements other than boron or lithium was ignored. Experiments were made with two similar detectors placed symmetrically on opposite sides of the paraffin with the absorber on one side only or, alternatively, with a detector on one side only both with and without the absorber. The β-activities of the detector were measured by means of an ionisation chamber fitted with an aluminium window 60 mm in diameter and 0.1 mm thick. The chamber was filled with argon at 3 atmospheres pressure and connected to an Edelmann electrometer. The ratio of detector activities with and without the absorber enabled the absorption cross-section of the absorber to be found, after allowing for the neutron angular distribution and the higher probability of capture of neutrons travelling obliquely.

An experiment was now made without the cadmium covering round the paraffin, with rhodium as the detector material and with a boron absorber giving a neutron transmission not far removed from 50 per cent. On the assumption that almost all the neutrons absorbed were thermal, and since boron was known to obey the $1/v$ law at thermal energies, it followed, as in Section 13.6, that the energy of the equivalent monoenergetic beam would be 0.025 eV. The cross-section of boron at this energy was thus determined. The experiment was then repeated with the cadmium covering replaced and a second boron cross-section obtained, corresponding to the energy for resonance capture in rhodium. By assuming the $1/v$ law to hold for boron at higher than thermal energies, the rhodium resonance energy could at once be found. Support for this assumption was obtained by repeating the investigation with lithium as absorber instead of boron, whereupon a second value for the rhodium resonance energy was arrived at that differed from the first by only 18 per cent. With the cadmium covering in place, measurements of boron cross-section were made with detectors of ten elements other than rhodium and their resonance energy levels thereby determined. Of these eleven determinations, seven were within a factor of 2 up or down on the currently accepted value of the lowest resonance energy for the element concerned, two were up by a much greater factor but within the energy range over which resonances are now known to occur whilst the remaining two were down by a much greater factor and were thus entirely misleading. The lowest resonance energy measured was about 1 eV, indicating that cadmium transmitted all neutrons exceeding some energy rather lower than this.

19.2 **Width of resonance energy band**

Concurrently with these determinations of resonance energy levels, an attempt was made by Amaldi and Fermi to establish the corresponding energy ranges over which resonance capture was maintained. Their source of slow neutrons was the paraffin cylinder with embedded radon–beryllium source as described at the beginning of Section 17.3. A thin sample of the element being investigated was placed between a detector of the same element and a cadmium filter of effective thickness 0.27 g/cm^2 that rested on the upper surface of the paraffin. After allowing for the neutron angular distribution, the saturation activity of the detector with and without the sample gave the absorption cross-section $\Sigma_{a_{res}}$ of the element for resonance neutrons. On repeating the experiment without the cadmium, the corresponding detector activities, less those previously obtained, gave the absorption cross-section $\Sigma_{a_{th}}$ of the element for thermal neutrons. A similar detector shielded by cadmium on both sides was now inserted at a particular position within the paraffin to acquire an activity A_{res} due to resonance neutrons. This activity was subtracted from that of the detector at the same position but shielded by cadmium on one side only to give an activity A_{th} due to thermal neutrons.

The results of these experiments were interpreted somewhat as follows. On the assumption that the paraffin was of sufficiently large dimensions, that neutrons were generated throughout it at a uniform rate q_o per unit volume and that resonance capture was sufficiently small, the flux per unit energy could be found from equation (185) as

$$\phi'(E) = q_0/(\Sigma_s)_E E$$

Since, in addition, the resonance capture cross-section was assumed to remain uniform over a sharply defined band of energy between an upper limit E_{max} and a lower limit E_{min}, it followed that

$$A_{res} \propto \Sigma_{a_{res}} q_0 \int_{E_{min}}^{E_{max}} \frac{dE/E}{(\Sigma_s)_E}$$

$$\propto \Sigma_{a_{res}} \frac{q_0}{\Sigma_s} \log (E_{max}/E_{min})$$

on ignoring the energy dependence of Σ_s the scattering cross-section for paraffin. The constant of proportionality is a function only of the geometrical arrangement and the efficiency of the ionisation chamber for the β-activity of the detector. The neutron flux producing activity A_{th} is given by the expression in Section 17.3 to which the activity A_2 is proportioned so that

$$A_{th} \propto \Sigma_{a_{th}} (q_{th}/\Sigma_a) \frac{2D/L}{c_z + 2(D/L)}$$

where Σ_a and L are respectively the absorption cross-section and diffusion length for paraffin, q_{th} is the rate of generation of thermal neutrons per unit volume and the constant of proportionality is the same as the previous one. Although q_{th} and q_0 are not necessarily the same at the same position within the paraffin, their integrals over the volume of the paraffin will be nearly equal so that, if D/L is sufficiently small,

$$\log (E_{max}/E_{min}) \approx \frac{(\Sigma_s/\Sigma_{a_{res}}) \int A_{res} \mathrm{d}V}{(1/2L\Sigma_{a_{th}}) \int A_{th} \mathrm{d}V}$$

From an expression such as this, Amaldi and Fermi determined the width of the resonance energy band for rhodium and silver.

Present-day information on resonance capture is obtained directly from the more recent determinations of cross-sections as functions of energy. Results for a selection of resonance absorbers are given in Figure 19.1.

19.3 Effective absorption cross-section

For neutrons of less than once-collided energies, the slowing-down behaviour has only been determined in the absence of absorption. An approximate allowance for the effects of resonance capture, however, may be made as described by Placzek. Absorption is assumed to arise only from resonance capture and to be therefore confined to narrow energy bands, or 'resonance bands'. The band at the highest energy will be referred to as the first, the band at the next highest energy as the second and so on. Between the $(k-1)_{th}$ and k_{th} bands, the slowing-down density will be a constant, written q_k. If the bands are separated by a sufficiently wide energy interval, the slowing down between them will become asymptotic so that, by analogy with equation (199),

$$(\Sigma_s)_E \phi'(E) = q_k/\xi E \qquad (204)$$

This result gives the asymptotic flux per unit energy between the $(k-1)_{th}$ and k_{th} bands. The flux per unit energy at energy E_k within the k_{th} band is given by equation (197) as

$$E_k(\Sigma_s + \Sigma_a)_{E_k}\phi'(E) = q_k - \int_{E_k}^{E_k'} (\Sigma_a)_{E_1}\phi'(E_1)\mathrm{d}E_1$$

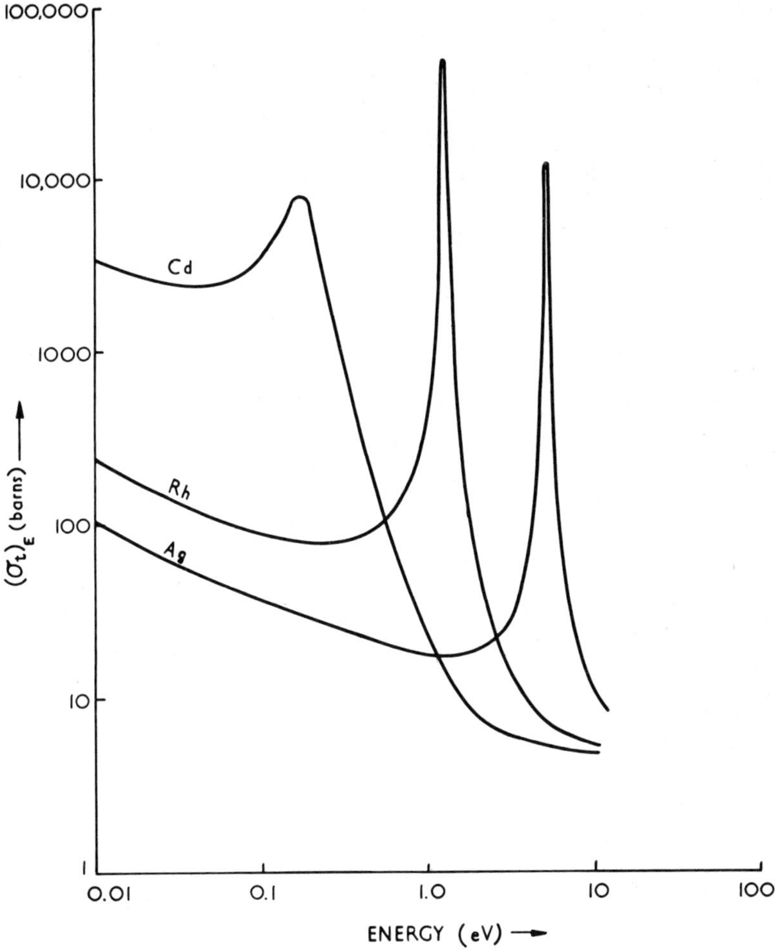

Figure 19.1 Selection of resonance absorbers

$$+ \frac{\alpha}{1-\alpha}\int_{E_k}^{E_k/\alpha} (\Sigma_s)_{E_1}\phi'(E_1)\mathrm{d}E_1$$

$$= q_k - \int_{E_k}^{E_k'} (\Sigma_a)_{E_1}\phi'(E_1)\mathrm{d}E_1 + \frac{\alpha}{1-\alpha}\int_{E_k}^{E_k'} (\Sigma_s)_{E_1}\phi'(E_1)\mathrm{d}E_1 + \frac{\alpha}{1-\alpha}\int_{E_k'}^{E_k/\alpha} (\Sigma_s)_{E_1}\phi'(E_1)\mathrm{d}E_1 \qquad (205)$$

where E_k' is the highest energy within the kth band. For achievement of asymptotic slowing down at energies just above E_k', the energy given by the upper limit on the third integral must be well below that of the $(k-1)_{\text{th}}$ band, and the integrand is therefore given by equation (204). Furthermore, if the reduction in slowing-down density over the k_{th} band is sufficiently small the first integral may be ignored in comparison with q_k. Again, if the k_{th} band is sufficiently narrow, the second integral may be ignored in comparison with the third so that

$$\begin{aligned} E_k(\Sigma_s + \Sigma_a)_{E_k}\phi'(E_k) &= q_k\left\{1 + \frac{\alpha/\xi}{1-\alpha}[\log(1/\alpha) - \log(E_k'/E_k)]\right\} \\ &= q_k/\xi \end{aligned} \tag{206}$$

if $\log(E_k'/E_k)$ is small in comparison with $\log(1/\alpha)$.

From equation (204) this result may be written in the form

$$(\Sigma_s + \Sigma_a)_{E_k}\phi'(E_k)\mathrm{d}E_k = (\Sigma_s)_{E_k}\phi'_{as}(E_k)\mathrm{d}E_k \tag{207}$$

where $\phi'_{as}(E_k)$ is the value at energy E_k of the asymptotic flux per unit energy above the k_{th} band. In this form the result is more readily given a physical interpretation. The right-hand side of equation (207) expresses the rate per unit volume at which neutrons are scattered out of the energy range $\mathrm{d}E_k$ in the absence of resonance capture. In such absence, however, this rate must also be the one at which neutrons are scattered into the energy range, and the rate of scattering in is practically the same whether there is resonance capture or not. It may thus be equated to the rate per unit volume at which neutrons are absorbed in and scattered out of the energy range $\mathrm{d}E_k$ and this rate is directly expressed by the left-hand side of equation (207).

It further follows from equation (207) that the flux per unit energy is depressed in a resonance band and bears a ratio to the asymptotic, or undepressed flux equal to

$$\left(\frac{\Sigma_s}{\Sigma_s + \Sigma_a}\right)_{E_k}$$

Sufficiently below the resonance band the flux per unit energy will continue to vary with energy as it did above the band apart from a reduction by the factor q_{k+1}/q_k. If E_k'' is the lower energy limit of the k_{th} band, then the rate of absorption in it is

$$q_k - q_{k+1} = \int_{E_k''}^{E_k'}(\Sigma_a)_{E_1}\phi'(E_1)\mathrm{d}E_1$$

$$= \int_{E_k''}^{E_k'} (\Sigma_{a_{eff}})_{E_1} \phi'_{as}(E_1) dE_1 \tag{208}$$

where, from equation (207),

$$\Sigma_{a_{eff}} = \Sigma_a \frac{\Sigma_s}{\Sigma_s + \Sigma_a} \tag{209}$$

is the 'effective absorption cross-section'.

As a result of flux depression in the resonance bands the absorption within a band remains finite, even when the absorption cross-section becomes infinite over a band of finite width. In this event the effective absorption cross-section in the band becomes merely Σ_s. On the other hand, when the absorption cross-section is sufficiently small, it becomes the limit approached by the effective cross-section.

More often than not, resonance capture at any particular energy level is attributable to one isotope only. Furthermore, in determining the macroscopic absorption cross-section of a material partly composed of such an isotope, it is more convenient to work with the absorption cross-section of the isotope in the microscopic form. Within a resonance band, the effective absorption cross-section in microscopic form is given by equation (209) as

$$\sigma_{a_{eff}} = \frac{\sigma_a}{1 + [\sigma_a/(\Sigma_s/N_{res})]} \tag{210}$$

where N_{res} is the number of atoms of the isotope per unit volume of the material.

The effective absorption cross-section is seen to depend only on the quantity Σ_s/N_{res}, which is the 'scattering cross-section per atom of absorber'. As this quantity becomes very large in relation to σ_a, the effective absorption cross-section increases to the actual one. As it becomes very small in relation to σ_a, the effective absorption cross-section falls to become equal to it. The effective absorption cross-section has its lowest possible value when the material consists of the resonance absorber isotope only. In this case, Σ_s/N_{res} will become equal to σ_s for the absorber. Since the value of σ_s is certain to be much less than that of σ_a, then it is the former value that will be approached by the effective absorption cross-section.

It remains to assess how great must be the energy interval between successive resonance bands for asymptotic slowing down to be achieved between them and how small must be the width of the bands themselves. The assessment of energy interval can be made by regarding each band as a negative source of neutrons. It then follows from Section 18.3 that the ratio of the energy levels of any two successive bands, or of the source and the band at highest energy, should be equal to $1/\alpha^2$ or more. An upper limit to the band width is set by the condition that the second integral in equation (205) should be small in comparison with the third, so that

$$E_k' - E_k'' \ll (E_k''/\alpha) - E_k'$$

or

$$(E_k'/E_k'') - 1 \ll \tfrac{1}{2}[(1/\alpha) - 1] \qquad (211)$$

A second upper limit is implied in the derivation of equation (206) and is set by the condition

$$\log(E_k'/E_k'') \ll \log(1/\alpha) \qquad (212)$$

This condition is satisfied, however, if the first is.

19.4 Resonance escape probability

The rate of absorption in the k_{th} resonance band can be obtained from equation (208) with the flux per unit energy as given by (204). In this way

$$q_k - q_{k+1} = \frac{q_k}{\xi}\int_{E_k''}^{E_k'}\left(\frac{\Sigma_a}{\Sigma_s + \Sigma_a}\right)_{E_1}\frac{dE_1}{E_1} \qquad (213)$$

a result derived by Placzek in 1940. The reduction in slowing-down density is thus proportional to the slowing-down density itself so that this result enables the fractional reduction to be obtained. Summing these reductions over all resonance bands

$$\sum_{k=0}^{k}\frac{q_k - q_{k+1}}{q_k} = \frac{1}{\xi}\int_{E_k''}^{E_0}\left(\frac{\Sigma_a}{\Sigma_s + \Sigma_a}\right)_{E_1}\frac{dE_1}{E_1}$$

the integrand being zero between successive bands. If $q_k - q_{k+1}$ is sufficiently small this result may be written

$$\int_q^{q_0}\frac{dq}{q} = \frac{1}{\xi}\int_E^{E_0}\left(\frac{\Sigma_a}{\Sigma_s + \Sigma_a}\right)_{E_1}\frac{dE_1}{E_1}$$

where q remains unchanged between successive resonance bands. On carrying out the integration,

$$q/q_0 = \exp\left[-\frac{1}{\xi}\int_E^{E_0}\left(\frac{\Sigma_a}{\Sigma_s + \Sigma_a}\right)_{E_1}\frac{dE_1}{E_1}\right] \qquad (214)$$

The right-hand side of this equation is termed the 'resonance escape probability' $p(E)$.

The resonance escape probability is very easily calculated for the case of a single resonance band such that the absorption cross-section bears a constant ratio to the scattering cross-section over the energy range $\frac{1}{2}\delta$ on either side of the resonance energy E_r but is otherwise zero. Equation (214) then gives the resonance escape probability as

$$p = \frac{1 - \dfrac{\delta/2}{E_r}}{1 + \dfrac{\delta/2}{E_r}} \quad \text{raised to the power } \frac{1}{\xi}\left(\frac{\Sigma_a}{\Sigma_s + \Sigma_a}\right)_{E_r}$$

$$= 1 - \left(\frac{\Sigma_a}{\Sigma_s + \Sigma_a}\right)_{E_r} \frac{\delta/E_r}{\xi} + \frac{1}{2}\left(\frac{\Sigma_a}{\Sigma_s + \Sigma_a}\right)_{E_r}^2 \left(\frac{\delta/E_r}{\xi}\right)^2$$

$$- \frac{1}{12}\left(\frac{\Sigma_a}{\Sigma_s + \Sigma_a}\right)_{E_r} \left\{\xi^2 + 2\left(\frac{\Sigma_a}{\Sigma_s + \Sigma_a}\right)_{E_r}^2\right\} \left(\frac{\delta/E_r}{\xi}\right)^3 + \ldots$$

Condition (211) may be satisfied by keeping $(\delta/E_r)/\xi$ small compared with unity. In this case, since the ratio of the second ξ^{-1} term to the first such term is equal to $(1/12)(\delta/E_r)^2$, p becomes very nearly a function of $[\Sigma_a/(\Sigma_s + \Sigma_a)]_{E_r} [(\delta/E_r)/\xi]$ as shown graphically in Figure 19.2.

On differentiating equation (214), the probability of a neutron undergoing resonance capture in slowing down through the energy range dE is found to be

$$dq/q = \left(\frac{\Sigma_a}{\Sigma_s + \Sigma_a}\right)_E \left(\frac{dE/E}{\xi}\right)$$

As pointed out at the end of Section 18.3, the second term in brackets on the right-hand side expresses the probability that the neutron will make a collision during asymptotic slowing down through the energy range dE. The other bracketed term expresses the probability that the collided neutron will then be absorbed. This term is seen to be proportional to $(\Sigma_a)_E$ when $(\Sigma_a)_E$ is small but to approach the limit unity as $(\Sigma_a)_E$ becomes very great. Over an energy interval within the resonance for which the absorption cross-section is sufficiently high, therefore, it is the logarithmic energy decrement relative to ξ that determines the extent of resonance capture rather than the absorption cross-section itself. This point is illustrated by Figure 19.2 for the special case to which it applies. On the other hand, if the logarithmic energy decrement over a resonance band is sufficiently great, the

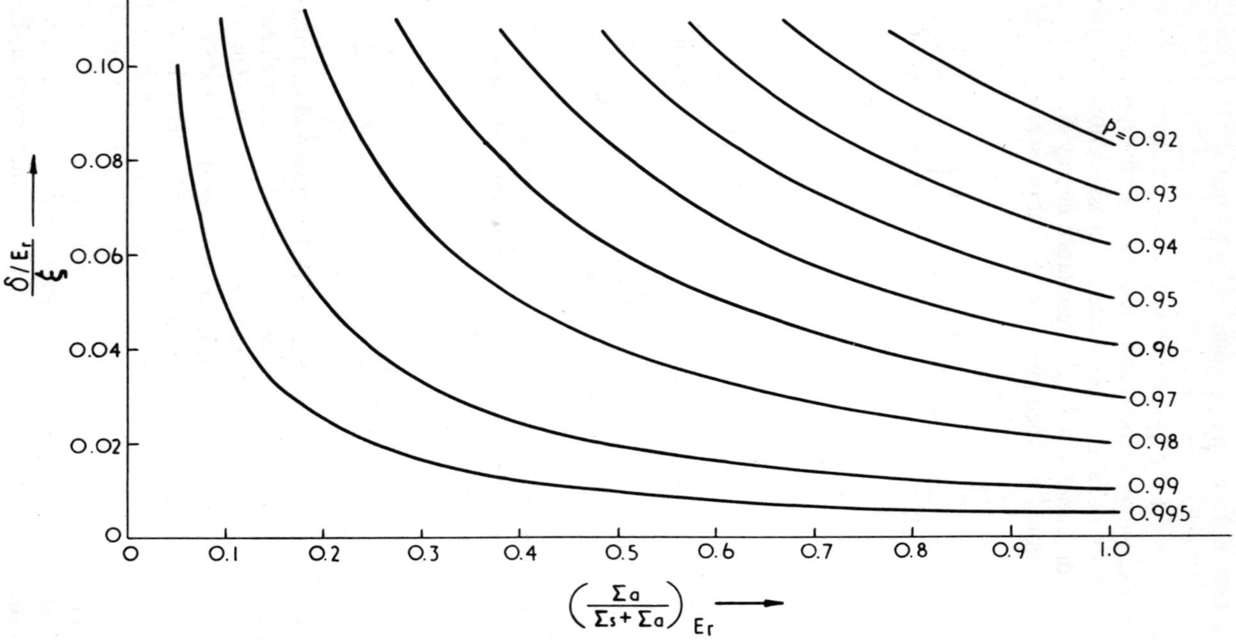

Figure 19.2 Resonance escape probability

condition (212), on which equation (214) is based, may be so far from being fulfilled that

$$E_k'/E_k'' \geq 1/\alpha$$

In this event, all neutrons just above energy E_k' will make a collision in the interval $E_k' - E_k''$ and it will then be the absorption cross-section that determines the extent of resonance capture. This situation cannot be covered by Figure 19.2, subject as it is to the condition (212). Nevertheless, it would not appear necessary to exceed the highest logarithmic energy decrement quoted in the figure for resonance capture to depend mainly on absorption cross-section.

At once-collided energies the effect of absorption on the slowing down of neutrons can be readily predicted from equation (184), whether the absorption takes the form of resonance capture or not. The resonance escape probability at these energies, however, is less easily formulated. An exception occurs when α is zero, as is very nearly true of hydrogen. In this event, introduction of equation (182) into (201) gives

$$\begin{aligned} q &= E(\Sigma_s + \Sigma_a)_E \phi'(E) \\ &= \left(\frac{\Sigma_s}{\Sigma_s + \Sigma_a}\right)_{E_0} q_0 \exp\left[-\int_E^{E_0}\left(\frac{\Sigma_a}{\Sigma_s + \Sigma_a}\right)_{E_1}\frac{dE_1}{E_1}\right] \end{aligned}$$

from (185). If absorption occurs only at energies less than E_0, then

$$\begin{aligned} p(E) &= q/q_0 \\ &= \exp\left[-\int_E^{E_0}\left(\frac{\Sigma_a}{\Sigma_s + \Sigma_a}\right)_{E_1}\frac{dE_1}{E_1}\right] \end{aligned}$$

Since ξ is unity when α is zero, it follows that equation (214) for resonance escape probability also applies exactly to the case of zero α.

When, as is generally true, resonance capture is attributable to one isotope only, it is convenient to express the resonance escape probability in microscopic form, as in the case of the effective absorption cross-section. Introducing equation (209) into (214),

$$p(E) = \exp\left[-\frac{1}{\xi\Sigma_s}\int_E^{E_0}(\Sigma_{a_{eff}})_{E_1}\frac{dE_1}{E_1}\right]$$

where Σ_s is the average scattering cross-section over the energy range E_0 to E. This result may be written

$$p(E) = \exp\left[-\frac{N_{res}}{\xi\Sigma_s}\int_E^{E_0}(\sigma_{a_{eff}})_{E_1}\frac{dE_1}{E_1}\right] \qquad (215)$$

where $\sigma_{a_{eff}}$ is as given by equation (210). The integral in equation (215) is called the 'effective resonance integral'.

19.5 Doppler effect

Although an absorption cross-section is generally quoted as a function of neutron energy it is, as pointed out in Section 11.2, more strictly dependent on the velocity of the neutron relative to the capturing nucleus. The distinction between the velocity of a neutron and its velocity relative to a capturing nucleus is normally of little account, but in the case of resonance capture it may be quite otherwise. Since the resonance energy is quoted in terms of a capturing nucleus at rest, a neutron at this energy will have a velocity relative to certain nuclei outside the range appropriate to resonance capture and the absorption cross-section for neutrons of resonance energy will thus be diminished. A neutron of energy just outside the resonance band, however, may yet have a velocity relative to certain nuclei within the range appropriate to resonance capture, which as a result will extend over a wider range of neutron energy. The influence on resonance capture of the velocity of the capturing nucleus is referred to as the 'Doppler effect' and was first recognised by H. A. Bethe and G. Placzek in 1936.

If a nucleus of mass number A has a velocity U, with a component U_n in the direction of motion of a neutron of velocity V, then from the equation (20) the collision energy will be

$$\epsilon = \tfrac{1}{2}M_n[A/(A + 1)]\,(V^2 + U^2 - 2VU_n)$$

where M_n is the mass of the neutron. Ignoring the term U^2 in comparison with V^2 and introducing the neutron energy

$$E = \tfrac{1}{2}M_n V^2$$

it is possible to write

$$\begin{aligned}\epsilon' &= [(A + 1)/A]\,\epsilon \\ &= E - (2M_n E)^{\frac{1}{2}}\,U_n \qquad (216)\end{aligned}$$

For nuclei of a given type, it is clear that the strict dependence of cross-section on ϵ implies an equal dependence on ϵ'. This dependence may be recognised by writing the cross-section as $(\sigma)_{\epsilon'}$. For the heavier nuclei the distinction between ϵ' and ϵ practically disappears.

The proportion of nuclei of mass M having a velocity component between U_n and $U_n + dU_n$ in any given direction follows the Maxwellian distribution

$$w(U_n)dU_n = (M/2\pi kT)^{\frac{1}{2}} \exp(-MU^2_n/2kT)dU_n$$

where T is the absolute temperature of the material containing the nucleus and k Boltzmann's constant. On introducing equation (216) this distribution may be written for the case of neutrons of fixed energy E as

$$w(\epsilon')d\epsilon' = \pi^{-\frac{1}{2}} \exp[-(E - \epsilon')^2/\Delta^2]d\epsilon'/\Delta \quad (217)$$

where

$$\Delta = 2(M_n E_r kT/M)^{\frac{1}{2}} \quad (218)$$

is the 'Doppler width'. In this last equation the resonance energy E_r has been written instead of E since the neutron energy is supposed to be at or near the resonance level. From the data of Appendix 1, equation (218) gives the Doppler width for nuclei of mass number A at a temperature of 20°C as

$$0.318(E_r/A)^{\frac{1}{2}} \quad (219)$$

where Δ and E_r are measured in electron-volts.

The cross-section for absorption of neutrons of energy E by nuclei of velocity such that the collision energy lies within the range $d\epsilon$ about ϵ is

$$w(\epsilon')(\sigma_a)_{\epsilon'}d\epsilon'$$

The absorption cross-section of the nuclei for neutrons of energy E is therefore

$$\begin{aligned}(\sigma_a)_{E,\Delta} &= \int_0^\infty w(\epsilon')(\sigma_a)_{\epsilon'}d\epsilon' \\ &= \pi^{-\frac{1}{2}}\int_0^\infty (\sigma_a)_{\epsilon'} \exp[-(E - \epsilon')^2/\Delta^2]\, d\epsilon'/\Delta \quad (220)\end{aligned}$$

from equation (217). It is seen to be a function not only of the neutron energy but also of the Doppler width, or effectively the temperature of the material.

In the simple case of resonance capture treated in Section 19.4, in which the absorption cross-section is equal to $(\sigma_a)_{E_r}$ at values of ϵ' between $E_r - \frac{1}{2}\delta$ and $E_r + \frac{1}{2}\delta$ and is zero otherwise, equation (220) becomes

$$(\sigma_a)_{E,\Delta}/(\sigma_a)_{E_r} = \pi^{-\frac{1}{2}}\int_{(E-E_r-\frac{1}{2}\delta)/\Delta}^{(E-E_r+\frac{1}{2}\delta)/\Delta} e^{-x^2}dx$$

$$= \frac{1}{2}\left\{\text{erf}\,\frac{(E-E_r+\frac{1}{2}\delta)/\delta}{\Delta/\delta} - \text{erf}\,\frac{(E-E_r-\frac{1}{2}\delta)/\delta}{\Delta/\delta}\right\}$$

where erf is the error function defined by $\text{erf}(x) = 2\pi^{-\frac{1}{2}}\int_0^x e^{-t^2}dt$.

From this result the ratio $(\sigma_a)_{E,\Delta}/(\sigma_a)_{E_r}$ has been plotted against $(E-E_r)/\delta$ for different values of Δ/δ in Figure 19.3. The Doppler effect is seen to be small when the Doppler width is less than 10 per cent of the width of the resonance band. With increasing Doppler width the resonance is extended over a greater energy range and at first the absorption cross-section continues to reach the same maximum value and half this value at the edges of the band. As the Doppler width increases above a quarter of the band width, the maximum value of the absorption cross-section falls but its value at the edges of the band remains unchanged. At Doppler widths above half the band width, however, the absorption cross-section at the edges of the band falls also. When the Doppler width has become equal to the band width the maximum value of the absorption cross-section is little more than half of that originally attained.

Inspection of Figure 19.3 suggests that the integral of resonance capture cross-section with respect to neutron energy is identical at all Doppler widths. This identity may be demonstrated not only for the simple case of resonance capture to which the figure applies but quite generally. The resonance capture cross-section is expressed as

$$(\sigma_a)_{\epsilon'} = \sum_m \sigma_{a_m}$$

where successive values of m differ by unity and

$$\sigma_{a_m} = (\sigma_a)_{m\delta\epsilon'} \text{ if } (m-\tfrac{1}{2})\delta\epsilon' < \epsilon' < (m+\tfrac{1}{2})\delta\epsilon'$$
$$= 0 \text{ otherwise.}$$

Equation (220) then becomes

$$(\sigma_a)_{E,\Delta} = \pi^{-\frac{1}{2}}\sum_m \sigma_{a_m}\int_{(m-\frac{1}{2})\sigma\epsilon'}^{(m+\frac{1}{2})\sigma\epsilon'} \exp\left[-(E-\epsilon')^2/\Delta^2\right] d\epsilon'/\Delta$$

$$= \pi^{-\frac{1}{2}}(\delta\epsilon'/\Delta)\sum_m \sigma_{a_m} \exp[-(E-m\delta\epsilon')^2/\Delta^2]$$

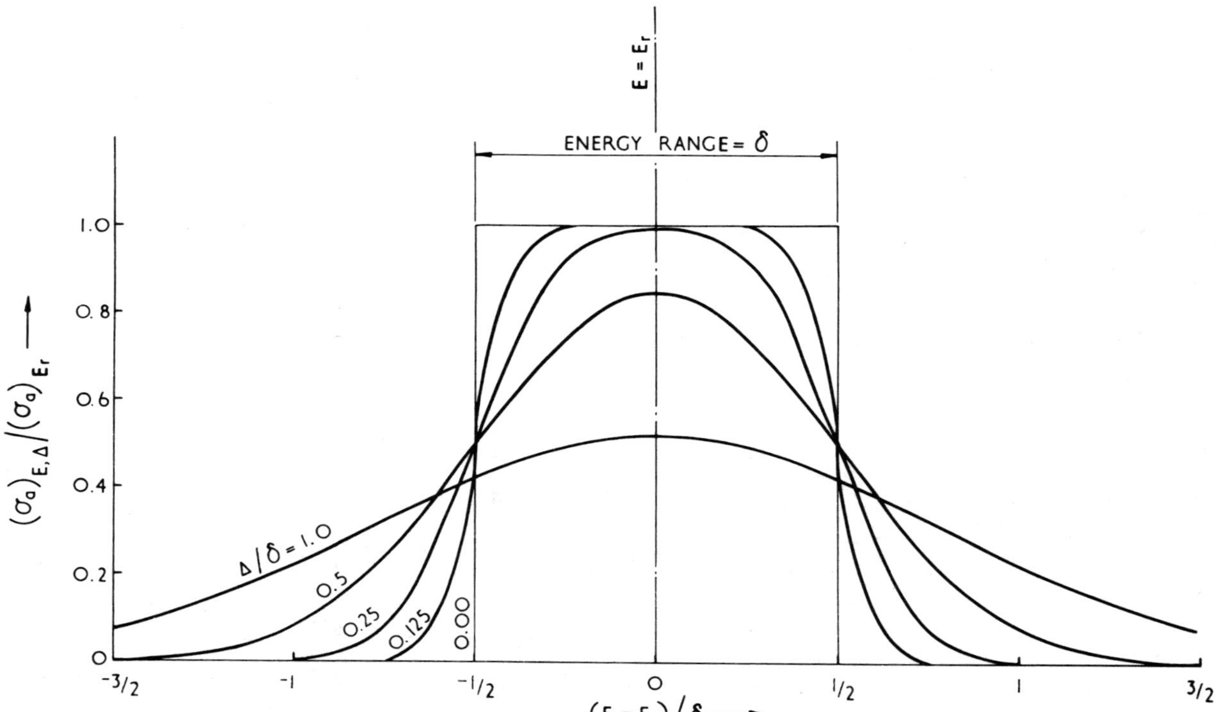

Figure 19.3 Doppler effect

if $\delta\epsilon'$ is sufficiently small. Setting $(E - m\delta\epsilon')/\Delta$ equal to x, then

$$\int_{-\infty}^{\infty} (\sigma_a)_{E,\Delta} \mathrm{d}E = \sum_m \left[\sigma_{a_m} \delta\epsilon' \pi^{-\frac{1}{2}} \int_{-\infty}^{\infty} \mathrm{e}^{-x^2} \mathrm{d}x \right]$$

$$= \sum_m \sigma_{a_m} \delta\epsilon'$$

$$= \int_{-\infty}^{\infty} (\sigma_a)_{\epsilon'} \mathrm{d}\epsilon'$$

in the limit when $\delta\epsilon'$ vanishes. This result, although mathematically rigorous, rests on the fiction of negative energies. In cases of practical interest, however, $(\sigma_a)_{E,\Delta}$ becomes vanishingly small well before E has fallen to zero and the lower limit $-\infty$ may therefore be replaced by zero with negligible error.

The extent to which the resonance escape probability is modified by the Doppler effect may be examined by first considering the case in which the absorption cross-section is not more than about 10 per cent of the scattering cross-section. Under these conditions equation (214) may be written

$$p \approx \exp\left[- \frac{1/\xi}{(\Sigma_s)_{E_r} E_r} \int_E^{E_0} (\Sigma_a)_{E_1} \mathrm{d}E_1 \right]$$

where $(\Sigma_s)_{E_r}$ is the scattering cross-section at the resonance energy E_r. Since the integral is unaffected by Doppler broadening of the resonance, it follows that the resonance escape probability is unaffected also. On the other hand, as mentioned in Section 19.4, over a narrow energy band in which the ratio of absorption to scattering cross-section is very large, the resonance escape probability becomes relatively insensitive to absorption cross-section but falls strongly with increase in band width. If the absorption cross-section is sufficiently high, therefore, its reduction on Doppler broadening will not only still leave the resonance absorption insensitive to it but will do so over a wider energy band and the resonance escape probability will thereby be diminished.

20

Slowing down of neutrons with diffusion

20.1 Diffusion equation for fast neutrons

Concurrently with the development of the slowing-down theory described in the previous two chapters, attempts were being made to remove its limitation to a source of infinite extent. The earliest of these were carried out by C. H. Westcott and E. Fermi with E. Amaldi and H. L. Anderson. Neutrons originating from a finite source will not only slow down but will diffuse as well. The spatial distribution of the neutrons may be described with the aid of a modified form of diffusion equation. It is the purpose of the present section to derive this equation.

An energy range will be considered having an upper limit of energy E_1 and a lower limit E_2. Neutrons slow down past energy E_1 with a slowing-down density q_1 and past energy E_2 with a slowing-down density q_2. On the assumption that they make a large number of scattering collisions in slowing down from energy E_1 to E_2, their behaviour in an element of volume $\mathrm{d}V$ is as shown schematically in Figure 20.1. The q_1 neutrons entering the element and those leaking into it by diffusion lose energy by successive scattering collisions, denoted by asterisks. Ultimately, they may be lost to the element by leakage or absorption. They may also be lost by slowing down out of the energy range and it is these neutrons that constitute the slowing-down density q_2.

It will first be supposed that all the q_1 neutrons slowing down into the element also slow down out of it. The number of collisions made by each such neutron in the process is then approximately

$$\log (E_1/E_2)/\xi$$

and the rate at which collisions occur in the element is therefore

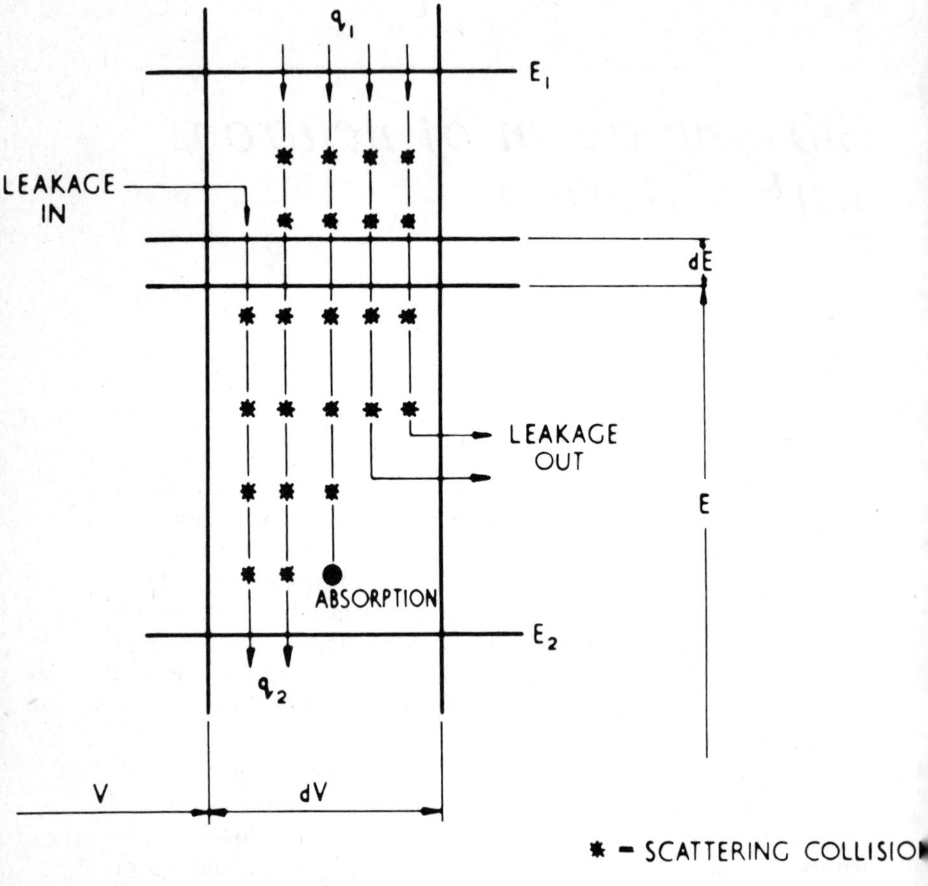

Figure 20.1 Group diffusion

$$S_s \, dV = q_1 \frac{\log (E_1/E_2)}{\xi} \, dV \tag{221}$$

This result may be corrected for the effect of absorption by subtracting from it the rate at which the absorbed neutrons would have collided had they continued to slow down from their energy at absorption to energy E_2. This rate is

$$S_a \, dV = dV \int_{E_2}^{E_1} (\Sigma_a)_E \phi'(E) \frac{\log (E/E_2)}{\xi} \, dE \tag{222}$$

A correction for leakage out of the element may be made by further subtracting the quantity

$$S_o\,\mathrm{d}V = -\,\mathrm{d}V\int_{E_2}^{E_1}(D_o)_E\nabla^2\phi'(E)\,\frac{\log\,(E/E_2)}{\xi}\,\mathrm{d}E \tag{223}$$

where $(D_o)_E$ has the value $(D)_E$ when $\nabla^2\phi'(E)$ is negative but is otherwise zero. Similarly, the neutrons leaking into the element will collide at a rate

$$S_i\,\mathrm{d}V = \mathrm{d}V\int_{E_2}^{E_1}(D_i)_E\nabla^2\phi'(E)\,\frac{\log\,(E/E_2)}{\xi}\,\mathrm{d}E \tag{224}$$

where $(D_i)_E$ now has the value $(D)_E$ when $\nabla^2\phi'(E)$ is positive but is otherwise zero. Finally, therefore, the total rate of collision in the element of volume $\mathrm{d}V$ is

$$(S_s - S_a - S_o + S_i)\,\mathrm{d}V = \mathrm{d}V\int_{E_2}^{E_1}(\Sigma_s)_E\phi'(E)\mathrm{d}E$$

On introducing equations (221) to (224) this result becomes

$$q_1\,\frac{\log\,(E_1/E_2)}{\xi} - \int_{E_2}^{E_1}(\Sigma_a)_E\phi'(E)\,\frac{\log\,(E/E_2)}{\xi}\,\mathrm{d}E + \int_{E_2}^{E_1}(D)_E\nabla^2\phi'(E)\,\frac{\log\,(E/E_2)}{\xi}\,\mathrm{d}E = \int_{E_2}^{E_1}(\Sigma_s)_E\,\phi'(E)\mathrm{d}E \tag{225}$$

This equation may be written more conveniently in terms of quantities averaged over the fast neutron energy range as

$$\left.\begin{aligned}
\phi_f &= \int_{E_2}^{E_1}\phi'(E)\mathrm{d}E\\
D_f\nabla^2\phi_f &= \int_{E_2}^{E_1}(D)_E\nabla^2\phi'(E)\mathrm{d}E\\
\Sigma_{a_f}\phi_f &= \int_{E_2}^{E_1}(\Sigma_a)_E\phi'(E)\mathrm{d}E\\
\Sigma_{s_f}\phi_f &= \int_{E_2}^{E_1}(\Sigma_s)_E\phi'(E)\mathrm{d}E\\
m &= \int_{E_2}^{E_1}(D)_E\nabla^2\phi'(E)\,\log\frac{E}{E_2}\,\mathrm{d}E/D_f\nabla^2\phi_f\log\,(E_1/E_2)\\
m' &= \int_{E_2}^{E_1}(\Sigma_a)_E\phi'(E)\,\log\frac{E}{E_2}\,\mathrm{d}E/\Sigma_{a_f}\phi_f\log\,(E_1/E_2)
\end{aligned}\right\} \tag{226}$$

It then follows that

$$q_1 - m'\Sigma_{a_f}\phi_f + m\,D_f\nabla^2\phi_f = \frac{\xi\Sigma_{s_f}}{\log\,(E_1/E_2)}\,\phi_f$$

A rearrangement of this equation gives

$$D_f \nabla^2 \phi_f - \Sigma_{a_f} \phi_f - \left[\frac{1}{m} \Sigma_{sl} \phi_f + \left(\frac{m'}{m} - 1 \right) \Sigma_{a_f} \phi_f - \left(\frac{1}{m} - 1 \right) q_1 \right] + q_1 = 0 \quad (227)$$

where

$$\Sigma_{sl} = \xi \, \Sigma_{s_f} / \log (E_1/E_2) \quad (228)$$

is the 'slowing-down' cross-section. The advantage of the rearrangement is that the term in square brackets represents the slowing-down density q_2. The fraction $\xi/\log (E_1/E_2)$ is the probability that a neutron making a scattering collision will leave the energy range $E_1 - E_2$, so that the slowing-down cross-section is the probability per unit distance of travel that a neutron will leave this energy range provided that absorption and diffusion are absent. In this case equation (227) gives $q_1 = q_2 = \Sigma_{sl} \phi_f$, which is (199) averaged over the range $E_1 - E_2$.

If the factor $(m'/m) - 1$ is small compared with unity the term containing it may be ignored so that we need only determine the parameter m. If the uniform slowing-down density that produces the same rate of collision per unit volume as that actually occurring is denoted by q_{av} then, from equations (221) and (226),

$$q_{av} \frac{\log (E_1/E_2)}{\xi} = \Sigma_{s_f} \phi_f \quad (229)$$

so that, from (228),

$$q_{av} = \Sigma_{sl} \phi_f$$

Since the term in square brackets in equation (227) is equal to q_2, it follows that

$$m = (q_1 - q_{av})/(q_1 - q_2) \quad (230)$$

As shown in Figure 20.2, m is thus a measure of the spectrum of slowing-down density. It has a high value when the slowing-down density changes rapidly with energy at high energies and a low value when the change is rapid at low energies.

Analogous to the diffusion length that characterises diffusion with absorption it is possible to define a parameter that characterises diffusion with slowing down. This parameter is termed the 'slowing-down length' and may be expressed as

$$L_s = \sqrt{(D/\Sigma_{sl})} \quad (231)$$

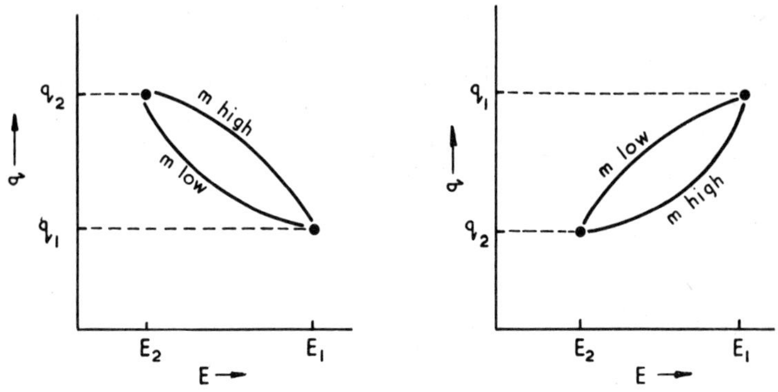

Figure 20.2 Spectrum of slowing-down density

Typical values for the slowing-down length are listed in Appendix 6.

As already noted, the diffusion equation (227) rests partly on the assumption that the number of collisions made by a neutron in passing through the energy range $E_1 - E_2$ is large compared to unity, and so cannot strictly apply to slowing-down media of light atoms. Moreover, if neutrons were to continue in the same direction with each collision, then the average distance travelled by them in slowing down over the energy range would be $\lambda_t \log (E_1/E_2)/\xi$, where λ_t is the total mean free path. But λ_t is of the order of the diffusion coefficient D so that this distance becomes of order $\Sigma_s L_s^2$. At some such distance from a localised neutron source the theory can be expected to break down.

A diffusion equation similar to (227) was developed by R. Ehrlich and H. Hurwitz Jr. in 1947 but based on an averaging procedure different from that used in (226) and with parameters analogous to m and m' but differently defined. Averaging procedure apart, the two diffusion equations can be made to agree when $m'/m = 1$ and m ranges from 0 to 1.

20.2 Fermi age theory

It is clear that the energy limits E_1, E_2 in the above treatment must be such that E_2/E_1 is less than α, the smallest fraction to which the energy of a neutron can be reduced in one scattering collision. When the mass number of the scattering nuclei is sufficiently large, however, α approaches unity and the energy limits E_1, E_2 may be made very close. If they are assumed to be infinitely close then ξ is required to be infinitesimal, as also is the scattering mean free path if $\xi\Sigma_s$ is to remain finite. In this way, the slowing-down process for a neutron is represented as an infinite number of collisions, each of which gives rise to an infinitesimal energy loss. On these assumptions it is convenient to write

$$\left.\begin{aligned} E_2 &= E \\ E_1 &= E + \mathrm{d}E \end{aligned}\right\} \qquad \left.\begin{aligned} q_2 &= q \\ q_1 &= q + \mathrm{d}q \end{aligned}\right\}$$

so that m approaches m' (with a common limit of ½) and equation (227) becomes

$$(D)_E \nabla^2 \phi'(E)\mathrm{d}E - (\Sigma_a)_E \phi'(E)\mathrm{d}E - \left[\frac{\xi(\Sigma_s)_E/m}{\mathrm{d}E/E} \phi'(E)\mathrm{d}E - \left(\frac{1}{m} - 1\right)(q + \mathrm{d}q) \right] + q + \mathrm{d}q = 0 \tag{232}$$

Now from equation (229)

$$\begin{aligned} \frac{\xi(\Sigma_s)_E}{\mathrm{d}E/E} \phi'(E)\mathrm{d}E &= q_{av} \\ &= (1 - m)(q + \mathrm{d}q) + mq \end{aligned}$$

from (230), so that

$$\phi'(E) = \frac{q + (1 - m)\mathrm{d}q}{\xi(\Sigma_s)_E E} \tag{233}$$

Introducing this result into equation (232)

$$\begin{aligned} &\nabla^2 \frac{(D)_E}{\xi(\Sigma_s)_E E} \left\{ q + (1 - m)\mathrm{d}q \right\} \mathrm{d}E \\ &= - \mathrm{d}q + \frac{(\Sigma_a)_E}{\xi(\Sigma_s)_E E} \left\{ q + (1 - m)\mathrm{d}q \right\} \mathrm{d}E \end{aligned} \tag{234}$$

Proceeding to the limit when $\mathrm{d}q$ is vanishingly small, equations (233) and (234) become

$$\phi'(E) = q/\xi(\Sigma_s)_E E \tag{235}$$

$$\nabla^2 q \frac{(D)_E}{\xi(\Sigma_s)_E E} dE = - \mathrm{d}q + q \frac{(\Sigma_a)_E}{\xi(\Sigma_s)_E E} \mathrm{d}E \tag{236}$$

Setting

$$q' = q \exp\left[\int_E^{E_0} \frac{(\Sigma_a)_E}{\xi(\Sigma_s)_E E} \mathrm{d}E \right] \tag{237}$$

$$\tau(E) = \int_E^{E_0} \frac{(D)_E}{\xi(\Sigma_s)_E E} \mathrm{d}E \tag{238}$$

equation (236) becomes, finally,

$$\nabla^2 q' = \partial q'/\partial\tau \tag{239}$$

In most slowing-down problems $(\Sigma_a)_E$ is negligible so that

$$q' = q$$

and equation (239) becomes

$$\nabla^2 q = \partial q/\partial\tau \tag{240}$$

If the diffusion coefficient and scattering cross-section are independent of energy then equation (238) reduces to

$$\tau(E) = (D/\xi\Sigma_s)\log(E_0/E) \tag{241}$$

and with E_0 replaced by E_1, $\tau(E_2)$ becomes identical to L_s^2 as given by (228) in (231).

These results constitute what is known as 'Fermi age theory'. The quantity τ is referred to as the 'age' of the neutrons and equation (240) as the 'Fermi age equation'. It was published by H. L. Anderson and E. Fermi in 1940. At the same time they gave the solution of the age equation for the case of a point source. The appropriate boundary conditions are

$$\left.\begin{array}{ll} & q = 0 \\ \text{when} & \tau = 0,\ r \neq 0 \end{array}\right\}$$

$$\text{and} \qquad 4\pi\int_0^\infty qr^2\,\mathrm{d}r = S$$

the source strength. The required solution is then

$$q = S\,\mathrm{e}^{-r^2/4\tau}/(4\pi\tau)^{3/2} \tag{242}$$

The ratio q/S is shown plotted against r for different values of τ in Figure 20.3. To some extent the scales are arbitrary. The same diagram is obtained if the scale of r is multiplied by any factor, provided that the values of τ are multiplied by the square of the factor and the scale of q/S divided by the cube of it.

For a given slowing-down medium, increasing values of the age τ correspond to lower neutron energies. At these energies neutrons will have undergone considerable slowing down and so diffused further from the source, as is indicated in Figure 20.3. For any given range of neutron energies, however, an increase in age could result from a smaller energy loss per collision or a smaller scattering cross-section. The first would lead to a large number of collisions and the second to a longer neutron mean free path so that, in either case, neutrons

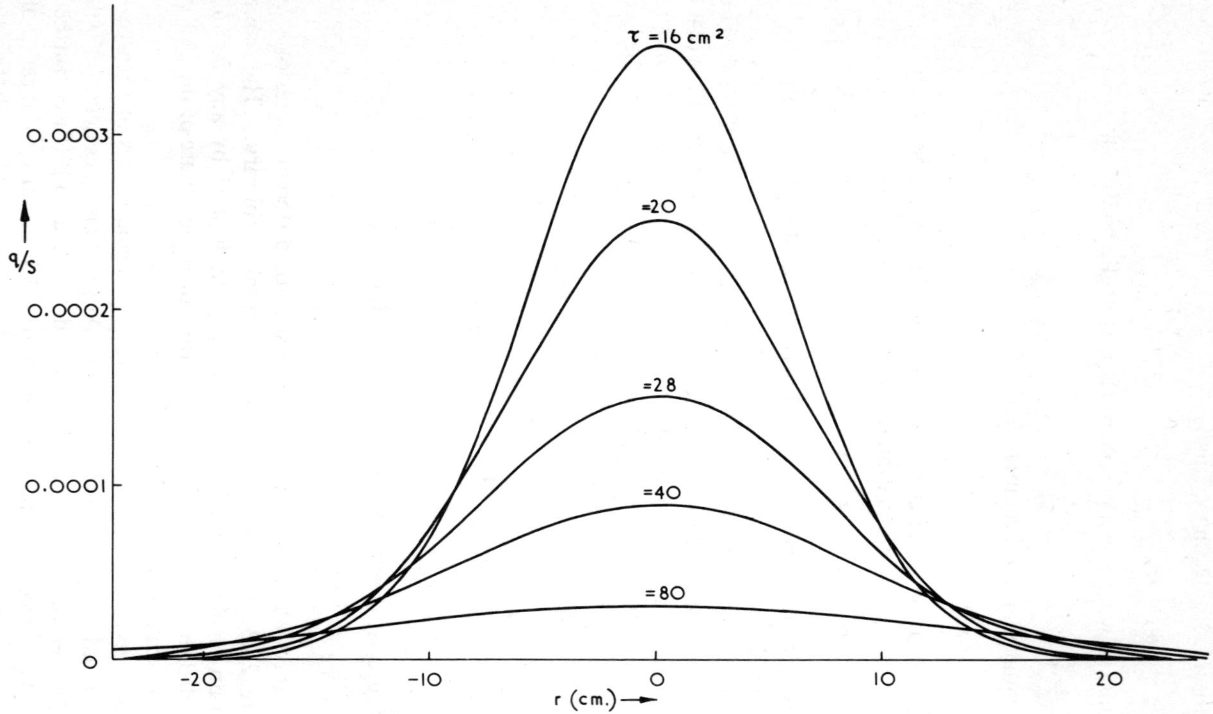

Figure 20.3 Slowing-down density with point source

would tend to have diffused further from the source, again as confirmed by the curves of the figure.

20.3 **Experimental support for age theory**

It has been seen that $\sqrt{\tau}$ is equal to the slowing-down length L_s. This length may be given a physical interpretation which has assumed some importance in the development of slowing-down theory. The mean square of the crow-flight distance travelled by a neutron from the source to a point at which its age is τ is

$$\begin{aligned}\overline{r}^2 &= 4\pi\int_0^\infty r^2\, qr^2\mathrm{d}r/4\pi\int_0^\infty qr^2\mathrm{d}r\\ &= 4\pi\int_0^\infty r^2\, qr^2\mathrm{d}r/S = 6\tau\end{aligned} \tag{243}$$

from equation (242).

If the neutron diffusion coefficient and scattering cross-section are independent of energy, therefore, the increase in $\overline{r}^2$ in slowing down from energy E_1 to E_2 is

$$\overline{r}_2^{\,2} - \overline{r}_1^{\,2} = 6(D/\xi\Sigma_s)\log(E_1/E_2) \tag{244}$$

For the case of hydrogen, this result was derived by Fermi in 1936. This derivation, however, was not obtained by applying age theory to a point source but from more fundamental considerations. It was carried out in two stages. In the first stage, neutrons were assumed to undergo a sequence of collisions, each associated with a prescribed angle of scatter, so that all slowing-down occurred through one and the same series of discrete energy levels. For any particular sequence of neutron paths, r^2 was obtained as their vector sum. This was now averaged over all possible lengths and directions of its constituent paths to give a mean value $\mathrm{r}_\mathrm{m}^{\,2}$. Proceeding to the second stage, the finite number of discrete energy levels was transformed into an infinite number of contiguous energy bands of infinitesimal width. In the resulting expression for $\mathrm{r}_\mathrm{m}^{\,2}$, the contribution from each energy band was weighted by the probability of a neutron colliding within it. The complete probability-weighted expression could now be integrated over the whole energy range $E_1 - E_2$ to give $\overline{r}^2$. The result was finally differentiated with respect to $\log_e (E_1/E_2)$ to obtain equation (244) in differential form.

An experimental check of the relationship (244) was reported by Amaldi and Fermi in 1936. A radon–beryllium neutron source was located in a tank of water of 900 mm diameter and 950 mm deep. Placed at various distances from this source was a detector, both source and detector being equidistant from the centre of the tank. A rhodium detector measuring 50 × 55 mm and of equivalent thickness 0.36 g/cm^2 was screened by cadmium of equivalent thickness 0.5 g/cm^2. The saturation activity was measured on both sides, the sum of the two activities being then proportional to the neutron flux at rhodium resonance energy. A similar determination was made without the cadmium so that the difference in the two gave the relative flux of thermal neutrons. The rhodium detector was next replaced by one of silver measuring 50 × 50 mm and of equivalent thickness 0.057 g/cm^2. With the cadmium screen in position the relative flux of neutrons of silver resonance energy could then be found.

For each of the three neutron energies, the relative neutron flux thus found was multiplied by r^2, where r was the distance between source and detector, to give a function $f(r)$. Then $\overline{r^2}$ was obtained as

$$\int_0^\infty r^2 f(r)\mathrm{d}r \Big/ \int_0^\infty f(r)\mathrm{d}r$$

The differences in the three values of $\overline{r^2}$ were then found, within limits of experimental error, to agree with those given by the relationship (244) from known values of the respective neutron energies, and the diffusion coefficient and scattering cross-section for water. The agreement was all the more remarkable in that age theory is least valid for very light atoms.

The use of age theory to interpret experiments on slowing down of neutrons in graphite was reported by Anderson and Fermi in 1940. A graphite stack 0.91 m square by 2.45 m in height was placed in the centre of a large room to minimise the effect of wall scattering. A radon–beryllium source was located at a height of 0.91 m near the vertical centre-line. Also close to the vertical centre-line was a vertical slot inside which a cadmium-screened rhodium or iodine detector could be placed in various positions. The rhodium detector was 50 mm square rhodium foil of effective thickness 0.126 g/cm^2. The iodine detector was a PbI_2 disc 60 mm in diameter and of effective thickness 0.35 g/cm^2.

The age equation (240) was solved to obtain a variation of q with distance z up the stack of the form $\exp(-z^2/4\tau)$ and the activity of the two detectors was found to approximate to a similar variation with appropriate values of τ. The difference between these values was within 20 per cent of that given by equation (241) from the known rhodium and iodine resonance energies, and the diffusion coefficient and scattering cross-section for graphite.

20.4 Correlation of fast neutron diffusion equation with age theory

Ignoring the term containing $(m'/m) - 1$, the fast neutron diffusion equation (227) may be written

$$D_f\nabla^2\phi_f - \Sigma_{a_f}\phi_f - (1/m)\Sigma_{sl}\phi_f + (q_1/m) = 0 \quad (245)$$

In the absence of absorption, this equation reduces to

$$\nabla^2\phi_f - (\phi_f/mL_s^2) + (q_1/mD_f) = 0$$

For the case of a point source of strength S this equation has the solution

$$\phi_f = (S/4\pi D_f r)\exp(-r/L_s\sqrt{m})$$

Also, since q_1 is zero at all points in the medium except at the source itself, equation (230) gives

$$q_2 = q_{av}/m$$

or, from (229),

$$q_2 = (\Sigma_{sl}/m)\phi_f$$
$$= (S/4\pi mL_s^2 r)\exp(-r/L_s\sqrt{m})$$

As r falls to zero this result leads to values of q_2 that approach infinity, and is thus strongly at variance with the age theory solution (242), whatever value is assigned to m.

Probably the best that can be done is so to adjust m that q_2 becomes small after a distance r that compares well with that given by age theory. The value of m selected may be tested against three alternative criteria. These are the achievement of the value given by age theory for

(i) $\overline{r^2}$, according to definition (243)

(ii) $\bar{r} = \int_0^\infty rq4\pi r^2\mathrm{d}r/S$

(iii) $r_{\frac{1}{2}}$ such that $\int_0^{r_{\frac{1}{2}}} q4\pi r^2\mathrm{d}r/S = \frac{1}{2}$

The values given by the two theories are listed in Table 20.1.

Table 20.1

Parameter	Age theory	Fast neutron diffusion equation
$\overline{r^2}$	$6\,\tau$	$6\,mL_s^2$
$\bar{r}$	$2.26\sqrt{\tau}$	$2.00\sqrt{m}L_s$
$r_{\frac{1}{2}}$	$2.18\sqrt{\tau}$	$1.68\sqrt{m}L_s$

If L_s^2 is to retain the value τ when the diffusion coefficient and scattering cross-section are independent of energy, then an appropriate value for m would appear to lie between 1.00 and 1.68. For values of m in this range, however, the value of q_2 given by the fast neutron diffusion equation will lie increasingly above that given by age theory as r falls below L_s and will lie increasingly below it as r exceeds about 1.5 L_s.

An alternative investigation into the value of m follows the lines of that developed by R. Ehrlich and H. Hurwitz Jun. in 1947 and published in 1954. The age theory solution in this case took the form

$$q' = \varrho(r)\eta(\tau)$$

so that the age equation (239) gave

$$\nabla^2\varrho + k^2\varrho = 0$$

and

$$q_2/q_1 = \mathrm{e}^{-x}$$

where

$$x = \int_{E_2}^{E_1}\left(\frac{\Sigma_a}{\xi\Sigma_s}\right)\frac{\mathrm{d}E}{E} + k^2(\tau_2 - \tau_1)$$

Eliminating ϕ_f in the diffusion equation (245) by means of (229), noting that if $\nabla^2\varrho + k^2\varrho = 0$ then $\nabla^2 q_{1,2} + k^2 q_{1,2} = 0$ so that $\nabla^2 q_{av} + k^2 q_{av} = 0$, and rewriting q_{av} in terms of q_1 and q_2 by means of (230), there results

$$q_1/m = [q_1 - m(q_1 - q_2)]\,[(\Sigma_a/\Sigma_{s1}) + k^2L_s^2 + (1/m)]$$

If the diffusion coefficient and scattering cross-section are independent of energy, then

$$(\Sigma_a/\Sigma_{s1}) + k^2L_s^2 = x$$

so that

$$q_2/q_1 = [1 - (1 - m)x]/(1 + mx)$$

With suffix a for the solution from age theory and suffix f for that from the fast neutron diffusion equation, it follows that

$$(q_2/q_1)_f = \frac{1 + (1 - m)\log(q_2/q_1)_a}{1 - m\log(q_2/q_1)_a}$$

The two solutions are compared in Table 20.2 for values of m equal to $\frac{1}{2}$ and 1.

Table 20.2

$(q_2/q_1)_a$	$(q_2/q_1)_f$	
	$m = ½$	$m = 1$
1.00	1.000	1.000
0.70	0.697	0.737
0.60	0.593	0.662
0.45	0.429	0.556
0.36	0.324	0.495
0.135	−0.001	0.333
0.010	−0.394	0.179
0.001	−0.551	0.127

Clearly the choice of $m = \frac{1}{2}$ leads to greater accuracy as q_2/q_1 approaches 1. However, even with $m = 1$, the two solutions agree to within less than 5 per cent provided that $(q_2/q_1)_a$ is greater than 0.7. But as $(q_2/q_1)_a$ falls below 0.6 the discrepancy between the two solutions exceeds 10 per cent. On the other hand, with $m = \frac{1}{2}$, agreement to within 5 per cent extends down to a value of $(q_2/q_1)_a$ = 0.45 and to within 10 per cent down to a value of 0.36. However, at values of $(q_2/q_1)_a$ less than 0.135, the corresponding values of $(q_2/q_1)_f$ become increasingly negative with this choice of m whilst with $m = 1$ they remain positive, although higher, and fall to zero when $(q_2/q_1)_a$ does so.

20.5 **Group diffusion theory**

As has already been seen, the fast neutron diffusion equation rests on the assumption that the average number of collisions made by a neutron between the source energy and the energy of interest must be large compared to unity. Also, the theory tends to break down at distances from the source of order greater than $\Sigma_s L_s^2$, although this limitation is of much less importance if the source is spatially distributed. These limitations also apply to age theory, the first of them even more strongly as the age equation corresponds to the fast neutron diffusion equation when its energy range is made infinitesimal. On the other hand, over the energy range to which it applies, the fast neutron diffusion equation makes only a crude allowance for the variation in neutron spectrum in contrast to the continuous spectrum incorporated in age theory. It is for this reason that age theory leads to more realistic results.

However, the main drawback to age theory is the difficulty of finding a solution that satisfies the boundary conditions between adjacent media. A simpler alternative is thus to subdivide the given energy range into a number of groups, in each of which the flux or slowing-down density is governed by an equation having boundary conditions that are simpler to apply. The obvious candidate for such an equation is the fast neutron diffusion equation (227) which now becomes a 'group diffusion equation'. The set of simultaneous group diffusion equations then covers the whole neutron spectrum from fast down to thermal, a representation known as 'group diffusion theory'.

In its simplest form the theory rests on two groups of neutrons, a fast group and a thermal group. As such it is referred to as 'two-group diffusion theory'. The equation for the fast group is simply the fast neutron diffusion equation (227) whilst that for the thermal group is (68) with the addition of a source term equal to the slowing-down density q_2 out of the fast group. With the approximation $m' = m$, the equations then take the form

$$D_1\nabla^2\phi_1 + q_1 - \left\{\frac{\Sigma_{sl}}{m}\phi_1 - \left(\frac{1}{m} - 1\right) q_1\right\} - \Sigma_{a_1}\phi_1 = 0$$

$$D_2\nabla^2\phi_2 + \left\{\frac{\Sigma_{sl}}{m}\phi_1 - \left(\frac{1}{m} - 1\right) q_1\right\} - \Sigma_{a_2}\phi_2 = 0$$

The representation of neutron slowing-down and diffusion can be further improved by increasing the number of energy groups, leading to 'multi-group diffusion theory'. The extension of the two-group diffusion equations to cover the multi-group case is straightforward and can be written as

$$D_1\nabla^2\phi_1 + q_1 - \left[\frac{\Sigma_{sl_1}}{m_1}\phi_1 - \left(\frac{1}{m_1} - 1\right) q_1\right] - \Sigma_{a_1}\phi_1 = 0$$

$$D_2\nabla^2\phi_2 + \left[\frac{\Sigma_{sl_1}}{m_1}\phi_1 - \left(\frac{1}{m_1} - 1\right) q_1\right] - \left[\frac{\Sigma_{sl_2}}{m_2}\phi_2\right.$$

$$- \left(\frac{1}{m_2} - 1\right) \left\{\frac{\Sigma_{sl_1}}{m_1}\phi_1 - \left(\frac{1}{m_1} - 1\right) q_1\right\}\Bigg] \quad - \Sigma_{a_2}\phi_2 = 0$$

$$D_3\nabla^2\phi_3 + \left[\frac{\Sigma_{sl_2}}{m_2}\phi_2 - \left(\frac{1}{m_2} - 1\right)\frac{\Sigma_{sl_1}}{m_1}\phi_1 + \left(\frac{1}{m_2} - 1\right)\left(\frac{1}{m_1} - 1\right) q_1\right]$$

$$- \left[\frac{\Sigma_{sl_3}}{m_3}\phi_3 - \left(\frac{1}{m_3} - 1\right)\left\{\frac{\Sigma_{sl_2}}{m_2}\phi_2 - \left(\frac{1}{m_2} - 1\right)\frac{\Sigma_{sl_1}}{m_1}\phi_1\right.\right.$$

$$\left.\left. + \left(\frac{1}{m_2} - 1\right)\left(\frac{1}{m_1} - 1\right) q_1\right\}\right] \quad - \Sigma_{a_3}\phi_3 = 0$$

$$\dots \quad \dots \quad \dots \quad \dots$$

where, in the last equation of the set, the third term representing neutrons slowing down out of the group is absent. As has already been seen, provided that the ratio of the slowing-down densities out of each group is not less than 70 per cent or so of the slowing-down density into it, then the choice $m = 1$ should lead to reasonable accuracy. At the same time this choice considerably simplifies the group diffusion equations, which now become

$$\left.\begin{array}{lllll} D_1\nabla^2\phi_1 & + q_1 & - \Sigma_{sl_1}\phi_1 & - \Sigma_{a_1}\phi_1 & = 0 \\ D_2\nabla^2\phi_2 & + \Sigma_{sl_1}\phi_1 & - \Sigma_{sl_2}\phi_2 & - \Sigma_{a_2}\phi_2 & = 0 \\ D_3\nabla^2\phi_3 & + \Sigma_{sl_2}\phi_2 & - \Sigma_{sl_3}\phi_3 & - \Sigma_{a_3}\phi_3 & = 0 \\ \dots & \dots & \dots & \dots & \\ D_n\nabla^2\phi_n & + \Sigma_{sl_{n-1}}\phi_{n-1} & & - \Sigma_{a_n}\phi_n & = 0 \end{array}\right\} \qquad (246)$$

As the number of groups is made larger the multi-group treatment becomes increasingly closer, as might be expected, to Fermi age theory. Not only, however, does the multi-group approach have the advantage of more tractable boundary conditions but, in forms where scattering across one or more groups is allowed for, it suffers less from some of the other limitations of age theory.

An early application of the idea of subdividing the neutron spectrum into discrete energy groups was made by Westcott in 1936. A point source emitting neutrons at rate S was assumed to be placed at the centre of a spherical cavity of radius r_0 in a medium of large extent. Three groups of neutrons were employed. At radius r, the nett current density of neutrons of the highest energy, or 'fast', group was taken to be

$$(S/4\pi r^2)\exp[-\Sigma_{sl}(r - r_0)]$$

The slowing-down density of neutrons into the middle, or 'intermediate', group was thereby determined. A neutron entering this group at radius r_1 was assumed to have a probability of leaving it within a radial element dr at r equal to $\psi(r - r_1)\mathrm{d}r$. On summing this probability for all neutrons entering the group between radii r_0 and r, the slowing-down density at radius r into the lowest, or 'slow', group was obtained. This slowing-down density replaced the ϕ_{n-1} term in the last of equations (246) to give the diffusion equation for the slow group.

The theory was tested against experiments made with a neutron source in a large vessel of water. Boron and silver detectors were used to measure the activity due to slow neutrons. The silver detector was also used with and without a 0.3 mm thick cadmium filter to obtain the activity due to thermal neutrons. The radius r_0 was 6 mm and known values were used for Σ_{sl} and the slow neutron diffusion coefficient and absorption cross-section. On taking

$$\psi(r - r_1) = g^2(r - r_1)\exp[-g(r - r_1)]$$

the variation in detector activity up to a radius of 130 mm could be made to agree closely with the calculated variation in slow neutron flux by suitable adjustment of the coefficient g.

The subdivision of neutrons into an intermediate energy group has become well established and intermediate neutrons are commonly assigned to energies ranging from about 0.1 MeV down to 10 eV or so. Energies above this range are then reserved for fast neutrons and those below it to slow neutrons.

21

Nuclear fission

21.1 Early ideas on activities of irradiated uranium

As already described in Section 10.4, the experiments carried out by Fermi and his co-workers in 1934 on the neutron irradiation of uranium had revealed radio-isotopes which had defied chemical identification. It was therefore natural that this problem should receive increasing attention in the next few years. One method of solving problems of this kind was described in Section 10.3. In that case the radio-isotope was identified by first dissolving in acid the material containing it and then adding to the solution stable compounds of elements of atomic number close to that of the target material. The radio-isotope was then deposited in association with a stable isotope of the same element. More generally the stable isotope, or 'carrier', can be any stable substance having chemical properties similar to those of the radio-isotope to be identified. A suitable carrier is thus often provided by a 'homologue' of the radio-isotope, that is a stable isotope that lies in the same subgroup of the Periodic Table. This technique was now about to be extensively used to telling effect by some of the most brilliant chemists the world has ever known.

Further experiments in which uranium was irradiated with unslowed neutrons were carried out by Lise Meitner, Otto Hahn and F. Strassmann in May 1937. They used a Geiger–Müller counter of aluminium with a wall thickness of 0.1 mm contained within a thick lead housing. The active substances took the form of finely divided precipitates pasted over with specially made thin gummed cellophane strips and were inserted into the counter housing on lead shuttles. By varying the irradiation time and analysing the decay curves as described in Section 5.4 the following decay series were proposed

$$\mathrm{U + n \longrightarrow {}_{92}U \xrightarrow{10\ s} {}_{93}EkaRe \xrightarrow{2.2\ m} {}_{94}EkaOs \xrightarrow{59\ m} {}_{95}EkaIr}$$

$$\mathrm{\xrightarrow{66\ h} {}_{96}EkaPt \xrightarrow{2.5\ h} {}_{97}EkaAu}$$

$$\mathrm{U + n \longrightarrow {}_{92}U \xrightarrow{40\ s} {}_{93}EkaRe \xrightarrow{16\ m} {}_{94}EkaOs \xrightarrow{57\ h} {}_{95}EkaIr}$$

where the successive elements in each series were lower homologues of the supposedly transuranic radio-isotopes.

The energy of the activating neutrons was examined with a hollowed-out paraffin cylinder, approximately 200 mm in diameter by 200 mm high, with a radon–beryllium source at its centre. A uranium preparation was placed within the cylinder with a 40 mm thickness of paraffin in front of it and a further thickness behind it. Its activity was found to be reduced by about 90 per cent on introducing a cadmium filter 0.42 mm thick on both sides of it, so that about 90 per cent of the activation was shown to be due to thermal neutrons. In order to find out if any of the remainder of the activation was due to uranium resonance neutrons, a thin detector was made up of a uranium preparation of effective thickness 67 mg U/cm^2. It was screened by a 0.46 mm thickness of cadmium on its face nearer the source to eliminate thermal neutrons and its activity measured. A second measurement was made with an absorbing layer of firmly pressed uranate of effective thickness 61 mg U/cm^2 inserted between the cadmium and the source. The difference in the two measurements gave the absorption of activating neutrons in the uranium resonance range. No such absorption was found, thus confirming that activation was due only to thermal and fast neutrons.

In addition to the activities mentioned above a further activity was found, having a half-life of 23 minutes. The experiment with the uranate absorbing layer was carried out as before to test whether the activity was due to uranium resonance neutrons. In this case the test proved strongly positive. Furthermore, the reduction in activity due to the absorbing layer showed its microscopic absorption cross-section to be about 1200 b(barns). A cross-section of this magnitude was unlikely to be attributable to the extremely small proportion of the uranium-235 isotope, and the resonance absorption was, therefore, attributed to uranium-238. By following the procedure adopted by H. H. Goldsmith and F. Rasetti, as described in Section 19.1, the energy of the uranium resonance neutrons was found to be about 25 eV. The decay series proposed for the activity was

$$\mathrm{U} + \mathrm{n} \longrightarrow {}_{92}\mathrm{U} \xrightarrow{23\ \mathrm{m}} {}_{93}\mathrm{EkaRe}$$

To account for the three decay series from the common uranium-239 parent, Hahn and Strassmann postulated that corresponding elements in the three decay series were isomers of one another and that β-decay occurred before isomeric transition could take place to a stable state. They admitted that such transition might occur beyond the radio-isotope EkaPt so that the three series could then merge into one. But they were also aware that the hypothesis raised difficulties and foresaw the possibility 'that the isomerism comes about on quite different grounds'.

21.2 First indications of sub-uranic active products

At about this time Irène Joliot-Curie and P. Savitch measured the activity of uranium after neutron irradiation and found activities having half-lives of 3.5 hours, 16 minutes, 2 minutes and about 40 seconds. Chemical tests showed the new 3.5-hour activity to be precipitated with lanthanum as a carrier. Since a disintegrating nucleus was so far known to change by only one or two electronic charges, the activity was attributed to an isotope of actinium, the homologue of lanthanum nearest in atomic number to uranium. However, the new activity was also precipitated with lanthanum from a solution containing both lanthanum and actinium. At this stage Curie and Savitch were not convinced that the activity could not be separated from lanthanum. Furthermore, having failed to trace its descent from any other active substance, they concluded that it was due to another transuranic radio-isotope.

In 1938 Hahn and Strassmann repeated these experiments and found activities having half-lives of 25 minutes, 110 minutes and several days. These activities were attributed to a radio-isotope of radium since they were precipitated with its lower homologue, barium. At the same time, the respective daughter activities were determined as 40 minutes, 4 hours and 60 hours. On the assumption that these were the products of β-decay, they were attributed to a radio-isotope of actinium. All these radio-isotopes were assumed to take on isomeric forms in order to account for the multiple decay chains. Furthermore, formation of the radium isomers was attributed to a rapid emission of two α-particles by the uranium-239 parent formed from neutron capture.

By the start of the following year, Hahn and Strassmann had found a fourth activity attributable to an isotope of radium having a half-life of <1 minute. They therefore revised their previous findings so that the radium isomers now became RaI, RaII, RaIII and RaIV having half-lives of <1 minute, 14 minutes, 86 minutes and 250–300 hours, respectively, whilst the actinium isomers became AcI, AcII, AcIII and AcIV having respective half-lives of <30 minutes, 2.5 hours, several days and <40 hours. However, on adding RaIV to a solution containing barium and an isotope of radium, the activity was found to associate with the barium rather than the radium. Hahn and Strassmann were thus driven to admit that 'as chemists we must in all truth say that, in the case of the new substances, we are dealing not with radium but with barium'. Likewise, on adding AcII to a solution of lanthanum and an isotope of actinium, subsequent chemical tests showed that AcII was not actinium, but at this stage its expected identity with lanthanum could not be wholly accepted.

By February 1939, Hahn and Strassmann had shown that their RaIII isotope was also an isotope of barium. Furthermore, its half-life of 86 minutes was almost identical to that of a β-emitting radio-barium, obtained from irradiating inactive barium with neutrons and presumed to have the mass number 139. Even after a 6-day irradiation, however, no activity could be detected from the daughter product of this radio-barium, as would be expected if it were the stable isotope lanthanum-139, so that the several day half-life of the 'AcIII' isotope appeared to be false. The RaIV isotope, already confirmed to be chemically identical with barium, was now suspected of being barium-140, since the half-life of its daughter product of <40 hours compared well with the known half-life of 31–46 hours for lanthanum-140.

These results led Hahn and Strassmann to consider whether the members of their previously proposed transuranic series might not be chemically identical to even lower homologues than those against which they had been chemically tested. These might therefore be radioactive forms of technetium (then known as masurium), ruthenium, etc. However, after consideration of these and other possibilities, the classification of the series as transuranic was maintained.

By now, Hahn and Strassmann were referring openly to 'bursting asunder' (*Zerplatzen*) of the uranium nucleus with emission of a few neutrons, and their active barium isotopes as one of the possible fragments. Uranium and barium having the respective atomic numbers 92 and 56, there remained a possible fragment of atomic number 36, i.e. krypton, with successive decay into rubidium, strontium, yttrium, etc.

Chemical tests on a solution of irradiated uranium mixed with strontium revealed the presence of active isotopes of strontium and yttrium together with a noble gas, though whether this was krypton, decaying to rubidium and strontium, or xenon, decaying to caesium and barium, could not at that time be confirmed.

21.3 Confirmation of nuclear fission

It was left to Lise Meitner and O. R. Frisch to propose the almost inevitable interpretation of these results, which they published in February 1939. Their proposal rested on the possibility that irradiation by neutrons caused the uranium nucleus to split into two roughly equal nuclei, the precise division probably depending partly on chance. Since the ratio of neutrons to protons in the nucleus generally increases with mass number it followed that the two nuclei resulting from the splitting, or 'fission', of the uranium nucleus would have an excess number of neutrons. This excess would then tend to be readjusted by β-decay. The barium isotopes, already found to be products of irradiated uranium, were now associated with one of the two nuclei formed by uranium fission. Associated with the second of these nuclei, therefore, were to be expected isotopes of atomic number equal to the difference between those of uranium and barium, that is isotopes of krypton, decaying perhaps through rubidium, etc. It now seemed probable that fission also accounted for the activities hitherto ascribed to members of the supposed transuranic series. In fact, chemical evidence suggested that the 10-minute and 40-second half-lives ascribed to uranium-239 in two of the series might be due to radioactive forms of technetium decaying through ruthenium, etc., as already suggested by Hahn and Strassmann. On the other hand, the member of the third series of 23-minute half-life had been chemically identified with uranium. It could therefore be confirmed as uranium-239, as originally proposed, rather than as a product of fission. The idea of nuclear fission now put forward removed, of course, any need to postulate nuclear isomerism to account for the multiple decay chains from irradiated uranium. Different activities of the same chemical element could now be attributed to different isotopes of the element due to differences in the proportions into which the uranium nucleus was split.

At the same time Meitner and Frisch quoted figures for the energy of the two nuclei, or 'fission fragments', resulting from uranium fission as about 200 McV, which they confirmed by means of the

Einstein mass–energy relationship. In view of the uncertainty attaching to the precise pair of fission fragments from any one fission, however, it is probably more convenient to apply this relationship by way of the concept of binding energy, as described in Section 9.3. From Figure 9.2, the binding energy per nucleon is seen to be about 7.6 MeV for uranium and about 8.5 MeV for fission fragments of mass numbers in the range 90–140. The energy released in the fission of uranium is thus about

$$238(8.5 - 7.6) = 214 \text{ MeV}$$

Meanwhile, Frisch reported the results of an experiment to confirm the generation of fission fragments from irradiated uranium. A uranium-lined ionisation chamber was connected to a linear amplifier and in turn to a thyratron biased to count only pulses exceeding 5×10^5 ion pairs. A neutron source consisting of 300 mg of radium mixed with beryllium was placed at a distance of 10 mm from the uranium lining. No counts were recorded with either the neutron source or the uranium lining removed. With the source and uranium lining in position, the maximum pulse was found to be at least 2×10^6 ion pairs. With the assumption of the same expenditure of energy in creating an ion pair as for α-particles, this pulse corresponded to the loss of 70 MeV in the energy of a particle in the chamber. Two such particles would therefore account for some 70 per cent of the energy release per fission, as estimated from the Einstein mass–energy relationship. Furthermore, the maximum possible distance of particle travel within the chamber was 30 mm in hydrogen at a pressure of 1 atmosphere, corresponding to 8 mm of air. The range of the particle was therefore likely to be small, as would be expected for the highly charged fission fragments. That the energy lost by the fission fragments through ionisation was less than that estimated from the Einstein mass–energy relationship was to be expected, since the energy lost in this way accounts only for the kinetic energy of the fission fragments and excludes other contributions such as those from radioactive decay of the fission products, that is, the fission fragments together with their daughter products. The energy released by fission of uranium is now generally accepted as 200 MeV.

21.4 **Fission product decay chains**

From now on the task of determining the various fission fragments and their decay chains was pursued on all sides with almost frenzied enthusiasm. By March 1939, Hahn and Strassmann in Germany had already demonstrated the production by uranium fission of barium, lanthanum, strontium, yttrium, xenon and caesium. In the same month F. A. Heyn, A. H. W. Aten Jun., and C. J. Bakker in Holland published evidence for the decay series

$$^{139}\mathrm{Xe} \xrightarrow{c.0.5\ \mathrm{m}} {}^{139}\mathrm{Cs} \xrightarrow{10\ \mathrm{m}} {}^{139}\mathrm{Ba} \xrightarrow{87\ \mathrm{m}} {}^{139}\mathrm{La}\ \text{(stable)}$$

$$^{88}\mathrm{Kr} \longrightarrow {}^{88}\mathrm{Rb} \xrightarrow{17\ \mathrm{m}} {}^{88}\mathrm{Sr}\ \text{(stable)}$$

and a month later E. Bretscher and C. G. Cook in England had assigned 4- and 40-hour activities to lanthanum and a 2.5-hour activity to bromine. In the following month P. Abelson in the USA recorded his proposals for the decay series

$$^{127}\mathrm{Sb} \xrightarrow{82\ \mathrm{h}} {}^{127}\mathrm{Te} \xrightarrow{10\ \mathrm{h}} {}^{127}\mathrm{I}$$

$$^{129}\mathrm{Sb} \xrightarrow{4.6\ \mathrm{h}} {}^{129}\mathrm{Te} \xrightarrow{70\ \mathrm{m}} {}^{129}\mathrm{I}$$

$$^{131}\mathrm{Te} \xrightarrow{30\ \mathrm{h}} {}^{131}\mathrm{I} \xrightarrow{8\ \mathrm{d}}$$

$$^{131}\mathrm{Te} \xrightarrow{25\ \mathrm{m}} {}^{131}\mathrm{I} \xrightarrow{8\ \mathrm{d}}$$

whilst by this time a 13-hour activity had been attributed to iodine-130 by J. J. Livingood and G. T. Seaborg. A selection of the various decay series from uranium fission as finally arrived at is given in Appendix 7.

The proportion of fissions that result in β-decay to the stable atom(s) of a given mass number is termed the 'chain yield'. Early work on yields was reported by H. L. Anderson, E. Fermi and A. V. Grosse in 1940. A solution containing 50 grams of uranium was irradiated inside a paraffin block near a cyclotron, which produced neutrons of up to 2 MeV in energy by bombardment of beryllium with 6-MeV protons. The intensity of irradiation was monitored by a gold foil placed inside the paraffin block. The arrangement was such that practically all fissions were produced by slow neutrons in uranium-235. It would appear, however, that the effect of neutrons emitted as a result of fission was not allowed for, although such emission was already being established, as described in the next chapter.

To obtain the rate of fission, the uranium solution was replaced by

140 cm^3 of M_nSO_4 solution, having a concentration such that the rates of neutron absorption in the two solutions were roughly the same, to give similar neutron distributions. Counts of the activity of a known fraction of the manganese atoms enabled the saturation activity of the whole solution to be deduced. This in turn could be factored by the ratio of uranium fission to absorption cross-sections, for which the first results were already to hand, as described below. The result was taken as the fission rate of the uranium solution and was used to calibrate the gold foil monitor.

To investigate a particular element occurring in a decay chain, a known amount of it was added as a carrier to the uranium solution. After irradiation it was purified and separated, and a weighed sample of it was painted on a thin aluminium strip, covered with Scotch cellulose tape and wrapped round a thin-walled silvered-glass counter. A correction factor η for the absorption of β-rays was obtained by comparing the count rate with and without an aluminium cylinder between sample and counter, the thickness of the cylinder being such as to give the same β-ray absorption as the counter wall and Scotch tape combined. The geometrical efficiency g of the arrangement was evaluated by counting the β-activity of a weighed amount of U_3O_8.

The weight of the sample of the carrier element offered to the counter as a fraction of the weight added to the uranium solution can be represented by f. Then, if the count rate were C, the activity of the whole of the carrier element added to the solution would be given by the expression

$$C\eta/gf$$

Using as carrier an element occurring in one or more decay chains, the activities were measured for various periods of irradiation and times of decay so that the corresponding saturation activities could be deduced. Division by the fission rate then gave the yield.

The element iodine was used as a carrier to obtain results for the decay chains of mass numbers 131 to 135, the element antimony for chains of mass numbers 127 and 129 and the element barium for the chain of mass number 140. These results were within a factor of 2 of present-day values, which are quoted in Appendix 7.

21.5 Neutron cross-sections of uranium

The possibility of fission introduces a new type of nuclear cross-section, the fission cross-section, σ_f. The absorption cross-section σ_a is therefore now made up of two components, the fission cross-section and the non-fission, or 'radiative capture', cross-section σ_c.

A value for the thermal neutron capture cross-section of uranium-238 was reported by H. Von Halban Jun., L. Kowarski and P. Savitch in 1939. It was found by comparing the 25-minute β-activity in uranium with the activity in a gold foil using the same source of slow neutrons under the same geometrical conditions. Irradiations were made through an absorber of effective thickness (1 g U + 0.2 g Au)/cm^2 to reduce the resonance component of the activating radiation, and the saturation activity was measured by a Geiger–Müller counter. Measurements were made both with and without a cadmium filter between the neutron source and the target material, so that their difference gave the activity I due to thermal neutrons. The measurements on uranium were complicated by the presence of activities other than the 25-minute one but these were removed as far as possible by chemical treatment after irradiation.

Denoting the capture cross-section of the activated material by Σ_c, and the volume of the whole of the target material by V, the rate of production of the active nuclei was proportional to $\Sigma_c V$. To allow for the absorption of β-rays within the target material, however, the constant of proportionality included the factor ϵ/M, where ϵ was the effective thickness in mass per unit area of a β-absorbing screen that factored the observed activity by some specified amount and M was the mass of the target material. In accordance with normal practice, the absorption of β-rays was assumed to depend primarily on the effective thickness of the screen rather than its material composition. The factor ϵ/M thus provided a measure of the reduction in observed activity due to β-ray absorption within the target material. Finally therefore

$$I \propto \frac{w\sigma_c}{\mathscr{A}}\epsilon$$

where w was the proportion by weight of the atoms of interest in the target material and $\mathscr{A}$ their atomic weight. The thicknesses ϵ were determined as those of a copper screen that halved the β-activity from both the uranium and gold target materials whilst the thermal neutron capture cross-section for gold was already known. The ratio of the activities I for the two target materials then gave the thermal neutron capture cross-section for uranium-238 as

$$\sigma_{c_u} = 1.3 \pm 0.45 \text{ b}$$

In the same series of experiments a determination was also made of the resonance escape probability for uranium. A gold plate detector surrounded by cadmium was activated by slow neutrons from a 300 mg radium–beryllium photo-neutron source in a 1-litre cube of paraffin. Before reaching the detector the slow neutrons passed through a 20 mm thickness of 1.6 molar solution of $UO_2 (NO_3)_2$. The solution was rich enough in uranium and of sufficient thickness for only neutrons of energy greater than the uranium resonance band at 25 eV to activate the detector. The experiment was then repeated with the same volume of 1.6 molar solution but of ammonium nitrate instead of $UO_2 (NO_3)_2$. In both cases the activities were measured by an electrometer and their ratio gave the required resonance escape probability. The value obtained was

$$p = 0.84 \pm 0.025$$

Measurements of the fission cross-section of uranium were reported by H. L. Anderson, E. T. Booth, J. R. Dunning, E. Fermi, G. N. Glasoe and F. G. Slack in March 1939. The interior of an ionisation chamber was coated with a layer of uranium oxide of thickness corresponding to 0.5 mm air equivalent so that all fissions could be observed. The neutrons reaching the chamber were precisely known by using a radon–beryllium source of known intensity in a standard position in a paraffin block. The number of fissions was measured both with and without a cadmium absorber so that both the fast and thermal neutron contributions could be obtained. The fission cross-sections were thus determined as

about 2 b for thermal neutrons
and 0.1 b for fast neutrons

These workers also reported a suggestion by N. H. D. Bohr that since neutron irradiation of uranium can lead to the formation of uranium-239, only uranium-235 gives rise to fission and not uranium-238. This statement is now known to be true but only in respect of thermal neutrons. A selection of more recent cross-sections for uranium is given in Appendix 8. These may be used with the fractional abundances of 0.712 per cent for uranium-235 and 99.282 per cent for uranium-238 to give cross-sections for natural uranium. The results are found to be within a factor of 2 or so of those due to Anderson and his co-workers as quoted above.

The daughter product of the 23-minute activity of uranium-239 is now known as neptunium and this in turn undergoes a β-decay to form plutonium-239 which is relatively stable. Successive radiative captures of neutrons then result in the formation of plutonium-240, -241 and -242. Data for plutonium is also given in Appendix 8.

22

The nuclear chain reaction

22.1 Neutrons emitted by fission

As pointed out by Meitner and Frisch, the fission fragments have an excess number of neutrons that can be eliminated by subsequent β-decay. An alternative mechanism for eliminating the excess neutrons was put forward in March 1939 by H. Von Halban Jun., F. Joliot and L. Kowarski. Their proposal was that some of the excess neutrons might be liberated directly, either during fission or later from the fission fragments. To check this proposal a radium–beryllium photo-neutron source was placed in a 1.6 molar aqueous solution of uranyl nitrate and also in a 1.6 molar aqueous solution of ammonium nitrate, the hydrogen contents of the two solutions differing by only 2 per cent. The saturation activity I of a dysprosium detector due to thermal neutrons was measured at various distances r from the source. For both solutions the results were expressed as a curve of Ir^2 against r. The rate of absorption of thermal neutrons per unit volume is proportional to $I\Sigma_a$, where Σ_a is the thermal neutron absorption cross-section of the solution. Hence the area S under the curve is proportional to the rate of production of thermal neutrons divided by Σ_a. If additional nuclei brought into solution do no more than increase absorption and affect the hydrogen content, then the resulting curve would be expected to be geometrically similar to the original one and to enclose a reduced area. However, the curves given by the two solutions could not be made to coincide by factoring their ordinates and abscissae whilst the curve for the uranyl nitrate solution was found to enclose a larger area. Evidence was thus provided that the uranium nuclei in this solution were generating neutrons.

The results of this experiment were analysed to obtain a figure for the number of neutrons emitted per uranium fission, but the present account is mainly based on an alternative form of analysis published by L. A. Turner in January 1940.

The thermal neutron absorption cross-section of the ammonium nitrate solution will be denoted by Σ_a and that of the uranyl nitrate solution by $\Sigma_a + \Delta\Sigma_a$. In the case of the uranyl nitrate solution, the point neutron source of strength Q is assumed to give rise to a uniformly distributed source pQ of thermal neutrons, where p is the resonance escape probability. Then, of the Q neutrons from the source, the number undergoing fission will be

$$pQ\Sigma_f/(\Sigma_a + \Delta\Sigma_a)$$

where Σ_f is the fission cross-section for thermal neutrons. These fissions result in the generation of

$$\nu pQ\Sigma_f/(\Sigma_a + \Delta\Sigma_a)$$

secondary neutrons, where ν is the number of neutrons emitted per fission. These in turn are thermalised to produce further fissions which generate further neutrons, and so on. The total number of neutrons generated in the uranyl nitrate solution from the Q source neutrons thus finally becomes

$$\Delta Q = Q\left[\frac{\nu p\Sigma_f}{\Sigma_a + \Delta\Sigma_a} + \left(\frac{\nu p\Sigma_f}{\Sigma_a + \Delta\Sigma_a}\right)^2 + \ldots\right]$$

whence

$$Q + \Delta Q = Q/\left(1 - \frac{\nu p\Sigma_f}{\Sigma_a + \Delta\Sigma_a}\right)$$

From the definition of S already given above, it follows that

$$\frac{S + \Delta S}{S} = \frac{p(Q + \Delta Q)/(\Sigma_a + \Delta\Sigma_a)}{Q/\Sigma_a}$$

so that

$$\frac{Q + \Delta Q}{Q} = \frac{1}{p}\left(\frac{S + \Delta S}{S}\right)\left(\frac{\Sigma_a + \Delta\Sigma_a}{\Sigma_a}\right)$$

$$= 1/\{1 - [\nu p\Sigma_f/(\Sigma_a + \Delta\Sigma_a)]\}$$

whence

$$\nu = \frac{\Sigma_a}{\Sigma_f}\left(\frac{\Sigma_a + \Delta\Sigma_a}{p\Sigma_a} - \frac{S}{S + \Delta S}\right)$$

The known concentrations and absorption cross-sections of hydrogen, nitrogen and oxygen in the ammonium nitrate solution gave

$$\Sigma_a = 0.019 \text{ cm}^{-1}$$

whilst the difference in hydrogen content of the two solutions gave a contribution to $\Delta\Sigma_a$ of

$$\Delta\Sigma_{a_H} = -0.0007 \text{ cm}^{-1}$$

The uranium fission cross-section of 2.0 b found by Anderson, Booth, Dunning, Fermi, Glasoe and Slack gave

$$\Sigma_f = 0.0020 \text{ cm}^{-1}$$

and, together with the uranium capture cross-section of 1.3 b due to von Halban, Kowarski and Savitch, gave a uranium contribution to $\Delta\Sigma_a$ of

$$\Delta\Sigma_{a_u} = 0.0033 \text{ cm}^{-1}$$

It then followed that

$$\Delta\Sigma_a = 0.0033 - 0.0007$$
$$= 0.0026 \text{ cm}^{-1}$$

For the uranium resonance escape probability p, the value of 0.84 found by von Halban, Kowarski and Savitch was used. Also the ratio $\Delta S/S$ was found to be 0.05 so that, finally,

$$\nu = \frac{0.019}{0.0020}\left(\frac{0.0216}{0.84 \times 0.019} - \frac{1}{1.05}\right)$$
$$= 3.8$$

as compared with the figure of 3.5 obtained by von Halban, Joliot and Kowarski, and also from the alternative form of analysis by Turner. More recent results are lower, as shown in Appendix 8.

The realisation that neutrons were emitted by the fission of uranium was of course of outstanding significance, since it opened up the possibility that these 'fission' neutrons might produce further fissions and so on. In this way a 'chain reaction' might take place throughout the uranium leading to the continuous release of enormous amounts of energy.

22.2 Energy of fission neutrons

It is implicit in the calculation described in the previous section that the energy of the fission neutrons will be higher than thermal. Hard on determining the number of neutrons released by fission, the same team of von Halban, Joliot and Kowarski went on to attempt a direct measurement of their energy and reported the results obtained in May of the same year. To carry out the measurement, an ionisation chamber containing oxygen at atmospheric pressure was inserted in a mass of uranium oxide. This was contained in a nearly cubical box measuring 90 mm × 90 mm × 80 mm which was surrounded by a thick layer of paraffin wax. Buried in the wax was a 300 g radium-beryllium photo-neutron source enclosed within a lead shield having a thickness of 50 mm in the direction of the ionisation chamber. The box could be screened on all sides with cadmium foil and pulses were recorded by the chamber with this screen in position or removed. These pulses at once gave the energy of oxygen nuclei projected by fast neutrons which had been liberated in the uranium as a result of fission by thermal neutrons. Due to the large number of accidental pulses produced by high energy γ-rays from the neutron source, only nuclei recoiling with an energy of at least 1.5 MeV could be taken into account. The frequency of the pulses was found to fall off rapidly between 1.5 and 2.5 MeV and more slowly between 2.5 and 3.7 MeV. The frequency of all pulses was roughly doubled when the cadmium screen was removed but recoils of about 2 MeV in energy were then notably more frequent. On the assumption of a head-on collision, as in Section 9.1, between neutron and oxygen nucleus, it followed that neutrons having an energy of at least 11 MeV are emitted by uranium irradiated with thermal neutrons. It was further concluded that this emission of fast neutrons was probably simultaneous with fission and would account for part of the difference noted in Section 21.3 between the total energy released by fission and the kinetic energy of the fission fragments. But the inability to detect neutrons of less than 7 MeV or so was unfortunate in that energies at this level or higher constitute only the high energy tail of the 'fission spectrum'. More recent measurements show that the neutrons liberated by fission have a most probable energy of about 0.7 MeV and an average energy of about 2 MeV.

22.3 Diffusion and transport theory for a multiplying medium

Following the demonstration of uranium fission, its possibilities as a source of useful energy at once became the subject of calculation. The theories of neutron diffusion and transport so far developed for a purely scattering and absorbing medium now had to be modified to cover a multiplying medium as well. As has mainly been the practice so far, attention will be concentrated on the case in which all neutrons belong to a single energy group. However, the treatment now has to take account of a distributed source of fast neutrons which are generated at the rate per unit volume $\nu\Sigma_f \phi$. If the energy of the group is the average energy of all neutrons, then it is only necessary to ensure that all neutron cross-sections conform to that energy. But if, as has often been the case hitherto, attention is confined to the thermal group, then the presence of neutrons generated by fission requires further consideration. Inevitably, the assumption will be made that such neutrons are thermalised instantly so that diffusion during slowing down is ignored. However, certain refinements can be introduced. For instance, some account can be taken of resonance absorption during slowing down on replacing $\nu\Sigma_f$ by $p\nu\Sigma_f$, where p is the resonance escape probability. A further factor could allow for the extra fissions by unslowed neutrons, but refinements of this kind have been largely outmoded by the more realistic representations that have become feasible with the advent of the electronic digital computer. Nevertheless, the one-group treatment is perhaps of value in providing a ready insight into neutron behaviour from which useful conclusions can be drawn, whilst at the same time leading to results of historic importance.

In order to modify the account already given of diffusion theory, it is first noted that the term S in equation (49) is replaced by $\nu\Sigma_f \phi$. The effect is the same as though S had been set equal to zero and Σ_a replaced by $\Sigma_a - \nu\Sigma_f$. Furthermore, the scattering cross-section Σ_s in expression (50) must now be replaced by $\Sigma_s + \nu\Sigma_f$. The total cross-section in expression (52), however, will remain unchanged as the sum of the cross-sections for scattering and absorption, which now includes fission. It follows that the results (49), (50) and (52) may be applied as before as though the source term S were zero on replacing

Σ_s by $\Sigma_s + \nu\Sigma_f$
Σ_a by $\Sigma_a - \nu\Sigma_f$

and equating

Σ_t to $\Sigma_s + \Sigma_a$

as before. This replacement was in fact incorporated in the integral equation set up by R. Peicrls in 1939, as referred to in Section 14.3. The same substitutions may now be made in equation (67) with S set equal to zero, and the diffusion equation (68) replaced by

$$\nabla^2\phi + B^2\phi = 0 \tag{247}$$

where B is a constant to be determined. Equation (69) thus becomes

$$(\Sigma_a - \nu\Sigma_f)\phi = (\Sigma_s + \nu\Sigma_f)\{-\tfrac{1}{3}(B/\Sigma_t)^2 + \tfrac{1}{5}(B/\Sigma_t)^4 - \tfrac{1}{7}(B/\Sigma_t)^6 + \ldots\}\phi$$

so that (70) is replaced by

$$B/\Sigma_t = \frac{\Sigma_s + \nu\Sigma_f}{\Sigma_t}\tan^{-1}(B/\Sigma_t) \tag{248}$$

which is the result as actually reported by A. H. Wilson in 1943. The remainder of the previous analysis may now be converted to a form appropriate to a multiplying medium by the further replacement of L by $1/\mathrm{i}B$. Equation (72) thus becomes

$$D = (\nu\Sigma_f - \Sigma_a)/B^2 \tag{249}$$

The condition that B^2 be positive, as required by equation (247), is thus that $\nu\Sigma_f$ be greater than Σ_a, so that neutrons are generated by fission more rapidly than they are absorbed.

In the case of one-dimensional transport theory, the differential scattering cross-section Σ_ψ includes a contribution from the neutrons emitted by fission. The expression

$$\Sigma_\psi\, \mathrm{d}\Omega\phi^*(z, \theta')\,\mathrm{d}\Omega'$$

thus represents the rate per unit volume at which neutrons are directed into the solid angle $\mathrm{d}\Omega$ about direction Ω as a result of scattering and fission capture of neutrons travelling within the solid angle $\mathrm{d}\Omega'$ about direction Ω', making an angle ψ with direction Ω. It follows that Σ_s is again replaced by $\Sigma_s + \nu\Sigma_f$ in equation (96) onwards. Also, in order that Σ_t shall remain equal to the sum of Σ_a and Σ_s, Σ_a must again be replaced by $\Sigma_a - \nu\Sigma_f$. Furthermore, since equations (99) and (100) take no account of fission neutrons, they can only remain relevant on condition that anisotropy is confined to scattering. This condition may be observed on replacing $\overline{\cos\psi}$ by an effective value

$$(\overline{\cos\psi})_{\mathrm{eff}} = \frac{\Sigma_s}{\Sigma_s + \nu\Sigma_f}\overline{\cos\psi}$$

The constant $\varkappa$ in equation (105) is now replaced by $\mathrm{i}B$ so that (106) becomes

$$\frac{d^2}{dz^2}\phi(z) + B^2\phi(z) = 0 \tag{250}$$

and the solutions $\phi_1(z)$, $\phi_2(z)$ correspond to the respective values $+\mathrm{i}B$ and $-\mathrm{i}B$ for $\sqrt{(-B^2)}$. As before, the diffusion length L is replaced from now on by $1/\mathrm{i}B$. The remainder of the analysis may thus be converted to a form appropriate to a multiplying medium by making exactly the same substitutions as were made in the similar conversion of diffusion theory. It then follows that equation (115) is replaced by

$$\frac{\tan^{-1}(B/\Sigma_t)}{B/(\Sigma_s + \nu\Sigma_f)} = \frac{1 + 3(\nu\Sigma_f - \Sigma_a)(\Sigma_s + \nu\Sigma_f)(\overline{\cos\psi}/B^2)}{1 + 3(\nu\Sigma_f - \Sigma_a)\,\Sigma_t(\overline{\cos\psi}/B^2)} \tag{251}$$

which is the result as actually reported by A. H. Wilson in 1944.

Equation (251) with (249) enables $D\Sigma_t$ to be plotted against $(\nu\Sigma_f - \Sigma_a)/\Sigma_t$ for various values of $\overline{\cos\psi}$, as shown for two cases in Figure 22.1. When $(\nu\Sigma_f - \Sigma_a)/\Sigma_t$ is zero, $D\Sigma_t$ is seen to have the same values as in Figure 14.5 when Σ_a/Σ_t is zero, but the variation of $D\Sigma_t$ with $(\nu\Sigma_f - \Sigma_a)/\Sigma_t$ is flatter than it was before with Σ_a/Σ_t.

Equation (131) is now transformed into

$$c_z = [1 + 3(\nu\Sigma_f - \Sigma_a)\Sigma_t(\overline{\cos\psi}/B^2)](\Sigma_s + \nu\Sigma_f)(\Sigma_t/B^2)\log[1 + (B/\Sigma_t)^2] - 3(\nu\Sigma_f - \Sigma_a)(\Sigma_s + \nu\Sigma_f)(\overline{\cos\psi}/B^2) \tag{252}$$

whilst (135) becomes

$$c_x = c_y = \frac{\Sigma_s + \nu\Sigma_f}{B}[2\{1 + 3(\nu\Sigma_f - \Sigma_a)\,\Sigma_t(\overline{\cos\psi}/B^2)\}\{\sqrt{[(\Sigma_t/B)^2 + 1]} - (\Sigma_t/B)\} - 3(\nu\Sigma_f - \Sigma_a)(\overline{\cos\psi}/B)] \tag{253}$$

The functions c_z and $c_{x,y}$ are shown plotted against $(\nu\Sigma_f - \Sigma_a)/\Sigma_t$ for two values of $\overline{\cos\psi}$ in Figure 22.2. It is seen that the function that previously increased with Σ_a/Σ_t now decreases with $(\nu\Sigma_f - \Sigma_a)/\Sigma_t$ and vice versa. As in the case of $D\Sigma_t$, the variation with $(\nu\Sigma_f - \Sigma_a)/\Sigma_t$ is much flatter than it was before with Σ_a/Σ_t. Also as in the case of $D\Sigma_t$, however, when $(\nu\Sigma_f - \Sigma_a)/\Sigma_t$ is zero, the functions retain the values that they had previously when Σ_a/Σ_t was zero.

Equation (138) for the angular flux $\phi^*(z,\theta)$ now becomes

$$\phi^*(z,\theta) = \frac{(1/4\pi)\,[(\Sigma_s + \nu\Sigma_f)/(\nu\Sigma_f - \Sigma_a)]}{(\Sigma_t/B)^2 + \cos^2\theta}\,[\phi(\Sigma_t/B^2)(\nu\Sigma_f - \Sigma_a)\{1 - 3\,\frac{\nu\Sigma_f - \Sigma_a}{\Sigma_t}\,\overline{\cos\psi}\cos^2\theta\} + J_z\{1 + 3(\Sigma_t/B^2)(\nu\Sigma_f - \Sigma_a)\,\overline{\cos\psi}\}\cos\theta]$$

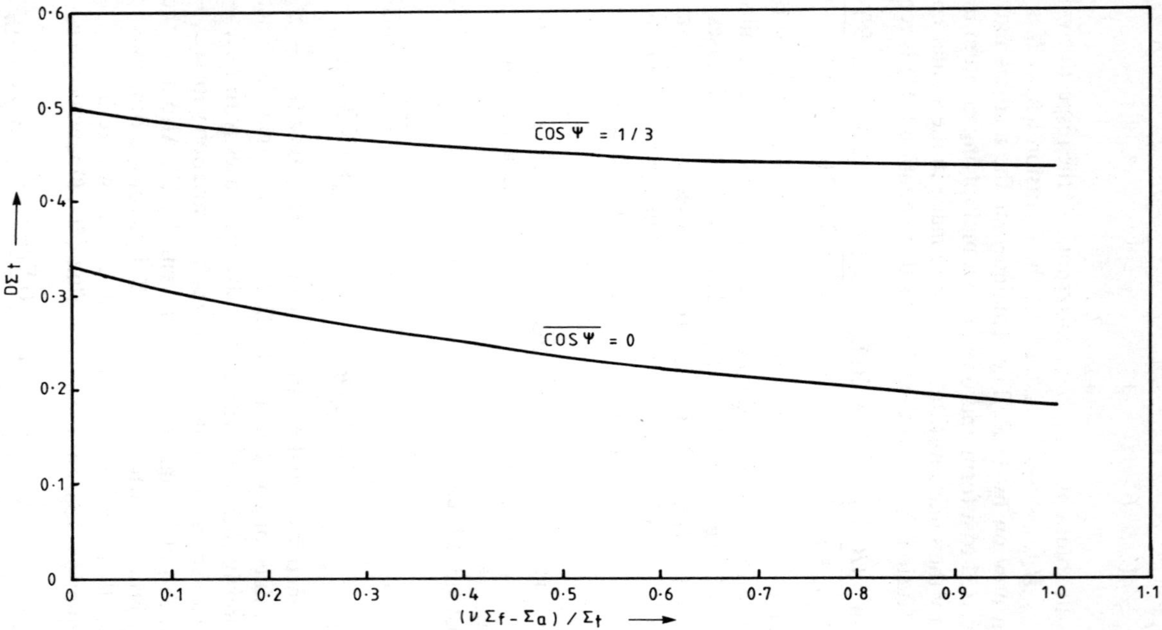

Figure 22.1 Diffusion coefficient for multiplying medium

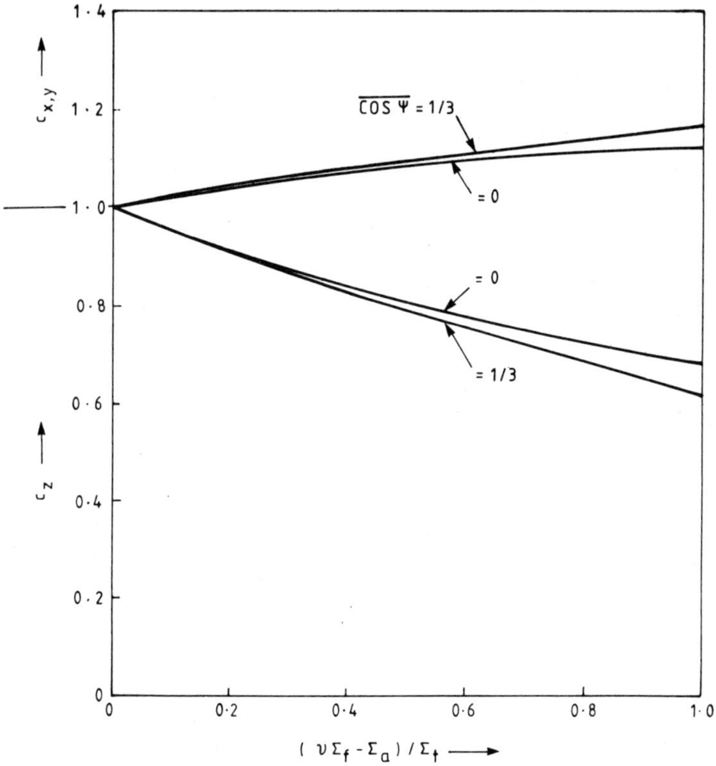

Figure 22.2 The functions c_z and $c_{x,y}$ for multiplying medium

When $(\nu\Sigma_f - \Sigma_a)$ approaches zero, the angular flux varies with θ as it did for a non-multiplying medium when Σ_a approached zero, as in the uppermost diagram of Figure 15.4. The effect of increasing $(\nu\Sigma_f - \Sigma_a)$ is shown in Figure 22.3 for two values of $\overline{\cos\psi}$. As compared with Figure 15.4, the curvature of the curves is generally reversed since neutrons that previously had the greatest chance of being absorbed now have the greatest chance of producing secondary fission neutrons.

The rules given in Section 17.1 for the application of diffusion theory can be followed as before, provided that equations (68) and (115) are replaced by (247) and (251) respectively. Finally, as applied to a multiplying medium, the group diffusion equations (246) also require to be modified. The modification can be arrived at on noting that equation (72) in (68) gives

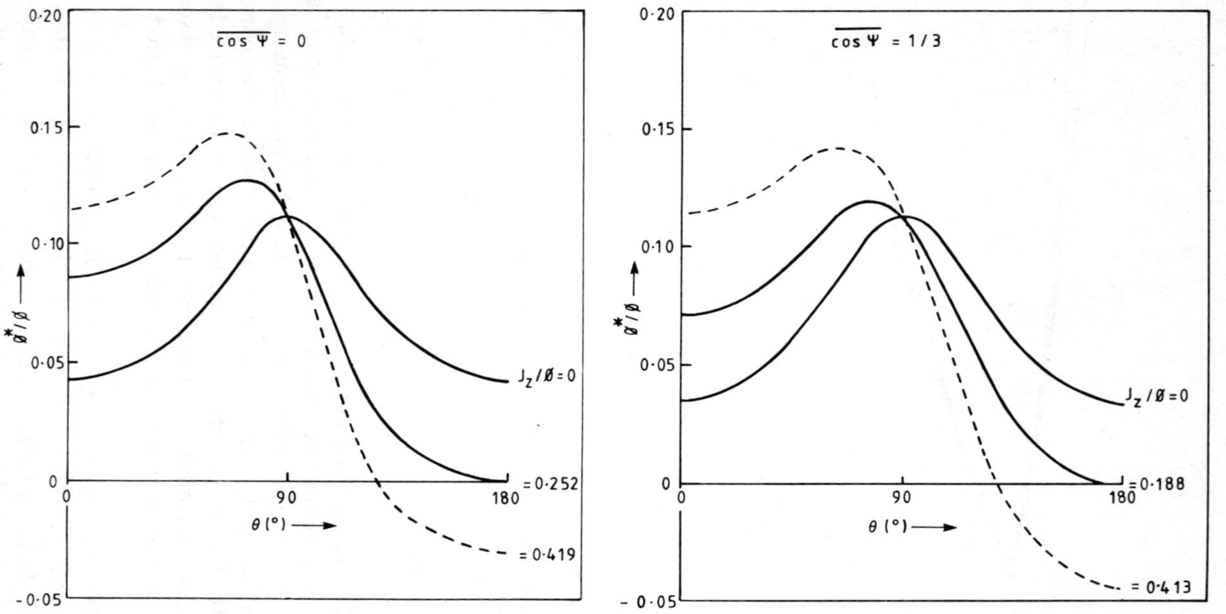

$(\nu \Sigma_f - \Sigma_a)/\Sigma_t = 0{\cdot}4$

Figure 22.3 Angular flux in a multiplying medium

$$D\nabla^2\phi - \Sigma_a\phi = 0$$

whilst (249) in (247) gives

$$D\nabla^2\phi - \Sigma_a\phi + \nu\Sigma_f\phi = 0$$

As might be expected, the modification takes the form of adding a neutron source term $\nu\Sigma_f\phi$ to the left-hand side of the previous diffusion equation. It follows that the group diffusion equations (246) are modified by including a term on the left-hand side equal to

$$+\sum_{i=1}^{n} \nu_i \Sigma_{f_i} \phi_i$$

which is factored by β_1 in the first equation, β_2 in the second, and so on. The summation represents the total rate at which neutrons are emitted by fission in all energy groups. The factor β_1 is then the proportion of these neutrons emitted with energies appropriate to the first group, the factor β_2 the proportion appropriate to the second, and so on.

22.4 Buckling and neutron flux

As pointed out in Section 14.4, the term $D\nabla^2\phi$ represents the net rate of neutron leakage into unit volume so that $B^2\phi$ in equation (247) is directly proportional to the net rate of neutron leakage out of unit volume. Since the neutron flux is inevitably positive, it follows that the spatial distribution of flux must be such that the flux gradient is continually decreasing as the flux level falls. In other words, the surface representing the distribution is convex, viewed from above. It further follows that the degree of convexity increases with the parameter B^2 so that, by analogy with the bowing of a strut, it is customary to refer to this parameter as the "buckling". By way of contrast, a similar line of reasoning shows that the surface representing the flux distribution in a purely absorbing medium is concave viewed from above and that the degree of concavity increases with the inverse square of the diffusion length.

Although a positive buckling is seen from the relation (249) to require that the rate of production of neutrons by fission exceeds the rate at which they are absorbed, this excess does not necessarily ensure a self-sustaining chain reaction. To explore the conditions under which such a reaction can take place, it is perhaps instructive to consider further the case of the slab of infinite face area with a neutron source uniformly distributed over an interior plane parallel to the faces, as discussed in Section 17.2. Taking as before the coordinate z to have its origin at the source plane, the flux ϕ for positive values of z is given by equation (174). Negative values of z are now represented by a new coordinate z' having the same origin as z and becoming increasingly positive as z becomes increasingly negative, as shown in Figure 22.4. In terms of the new coordinate the neutron flux will be written ϕ'. The complete solution for the flux can then be expressed as

$$\left.\begin{aligned} \phi &= \phi_0 \sinh \frac{T - z}{L} \Big/ \sinh \frac{T}{L} \\ \phi' &= \phi_0 \sinh \frac{T' - z'}{L} \Big/ \sinh \frac{T'}{L} \end{aligned}\right\} \tag{254}$$

where

$$\phi_0 = \frac{2L/D}{\coth \dfrac{T}{L} + \coth \dfrac{T'}{L}} C \tag{255}$$

in which C represents the neutron source strength previously denoted by C'. The form of the flux distribution typically obtained from equations (254) is illustrated by the uppermost curve in Figure 22.4. As pointed out in Section 17.2, the discontinuity in the flux gradient at the plane $z = 0$ corresponds to the jump $2C$ in the net current density. If it is desired to maintain the flux ϕ_0 at some given level then, as can be seen from the relation (255), the source strength C must be reduced as absorption falls.

As absorption continues to fall or, in the case of a multiplying medium, fission occurs on an increasing scale, a stage is eventually reached at which, in the first case, absorption has become zero or, in the second, the rate at which neutrons are generated by fission has become exactly equal to that at which they are absorbed. In the first case L will approach infinity and equations (254) and (255) reduce to

$$\left.\begin{aligned} \phi &= \phi_0 \left(1 - \frac{z}{T}\right) \\ \phi' &= \phi_0 \left(1 - \frac{z'}{T'}\right) \end{aligned}\right\} \tag{256}$$

where

$$\phi_0 = \frac{2/D}{(1/T) + (1/T')} C \tag{257}$$

To examine the second case it is first necessary to derive equations analogous to (254) and (255) for conditions under which B^2 is positive.

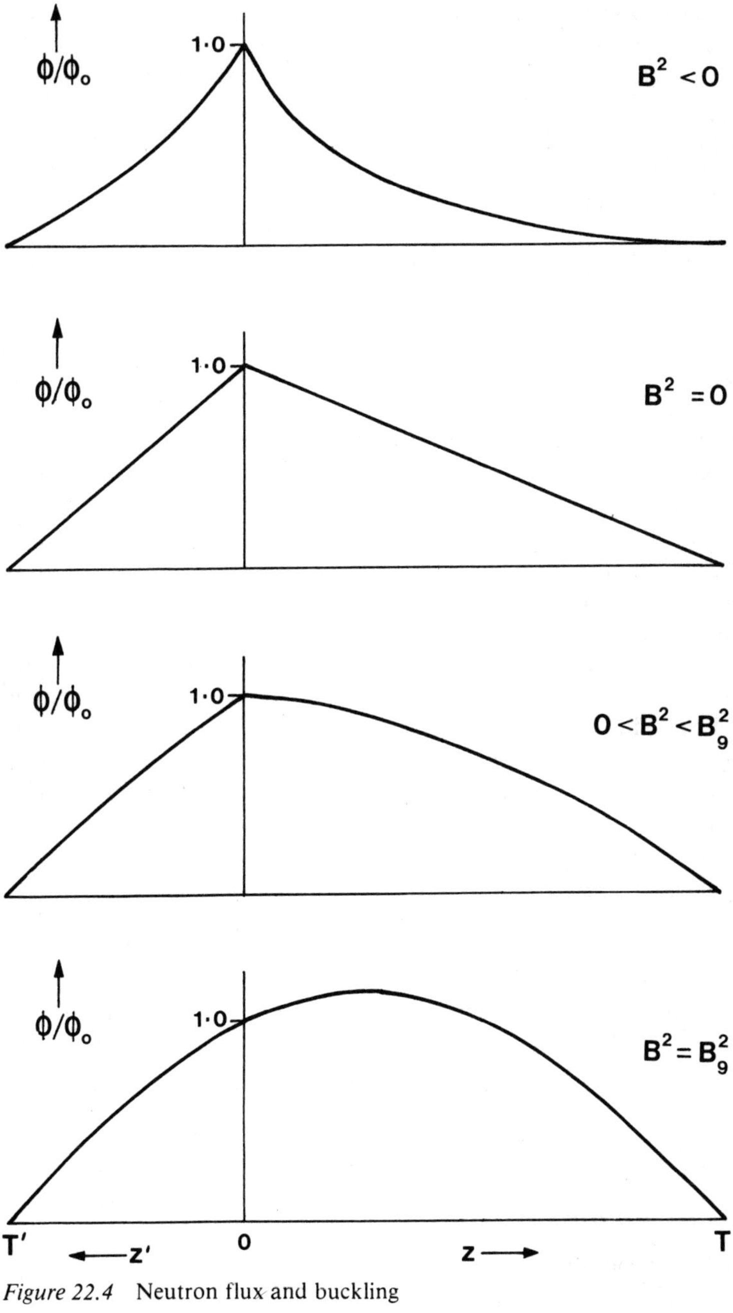

Figure 22.4 Neutron flux and buckling

The derivation may be made either by direct solution of the appropriate diffusion equation or by introducing the substitutions for Σ_s, Σ_a and L given in Section 22.3, together with the identities

$$\left.\begin{aligned} \sinh ix &= \mathrm{i} \sin x \\ \cosh ix &= \cos x \end{aligned}\right\}$$

The result obtained is

$$\left.\begin{aligned} \phi &= \phi_0 \sin B(T - z)/\sin BT \\ \phi' &= \phi_0 \sin B(T' - z')/\sin BT' \end{aligned}\right\} \quad (258)$$

where

$$\phi_0 = \frac{1/DB}{\cot BT + \cot BT'} C \quad (259)$$

For the case in which B^2 is zero, these equations reduce, as they should, to (256) and (257).

Whether absorption has fallen to zero in a purely absorbing medium, therefore, or fission has occurred on a scale sufficient to offset it in a multiplying one, the source neutrons are entirely accounted for by leakage and the flux falls off linearly from source to boundary as shown by the second curve in Figure 22.4.

As the buckling B^2 becomes increasingly positive, the curve representing the flux distribution becomes more convex, as shown by the third curve in Figure 22.4. Also, if the neutron flux at the source plane is to be maintained at the given level ϕ_0, then equation (259) shows that the source strength C must be even further reduced. This reduction is evident in Figure 22.4 from the smaller discontinuity in the flux gradient at the source plane.

22.5 Criticality

The preceding argument reaches its conclusion when, as B^2 increases still further, a stage is reached at which the flux level ϕ_0 can only be maintained unchanged if the source strength C vanishes. It follows that the flux gradient then loses its discontinuity at the source plane and becomes continuous throughout, as shown by the lowest curve in Figure 22.4. It further follows from the relation (259) that the coefficient of C must now approach infinity, so that

$$\cot BT + \cot BT' = 0$$

This condition is satisfied if

$$\sin B(T + T') = 0$$

i.e., if

$$B = \pi/(T + T'),\ 2\pi/(T + T'),\ \text{etc.} \tag{260}$$

When the buckling attains the lowest value given by the condition (260), therefore, the neutron flux can be maintained at any prescribed level without the aid of a neutron source introduced from outside. Since this value of the buckling is determined by the geometry of the system, it is referred to as the 'geometric' buckling, and will be denoted by B_g^2. In order to distinguish between it and the buckling B^2 determined from the material properties of the system, as by relations such as (249), the latter is sometimes referred to as the 'material' buckling. The system is now said to have become 'critical'. Any excess in buckling above the geometric one will cause the neutron flux to increase without limit so long as the excess is maintained. The system is then said to have become 'supercritical'. In a controlled reaction the system is only allowed to remain supercritical until the desired neutron flux level has been attained. The flux may then be held at that level by reintroducing absorber material to an extent sufficient to restore criticality. Likewise, to reduce the flux level, an excess of absorber is introduced so that the neutron flux falls, but at a continually decreasing rate. The system is then said to be 'subcritical'. When the neutron flux has fallen to the level required, it may again be held at that level by removing the excess absorber so that criticality is once more restored.

23

The nuclear reactor

23.1 Critical mass and reflector

The critical system in the form of a slab of infinite face area constitutes a hypothetical version of the nuclear reactor known as the 'slab' reactor. Although resembling no reactor in practice, it can nevertheless provide a simple means of predicting the size required to ensure criticality, or 'critical size', and so lead to useful conclusions on features which might affect it. A closer approach to reality is obtained on replacing the slab by a sphere to give the 'spherical' reactor. One of the earliest calculations on a spherical reactor to predict effects on critical size was that reported by Francis Perrin in May 1939, before the publication by Peierls of his integral equation in the same year. For this form of reactor the diffusion equation for the flux ϕ at radius r becomes

$$\frac{d^2\phi}{dr^2} + (2/r)\frac{d\phi}{dr} + B^2\phi = 0 \qquad (261)$$

In order that the neutron flux should vanish at the extrapolated outside surface of radius R, the solution took the form

$$\phi \propto (1/r) \sin B(R - r)$$

The condition that the flux should remain finite at the centre was then

$$B = \pi/R$$

so that $(\pi/R)^2$ is the geometric buckling for this case. Absorption in atoms other than those of uranium was ignored so that the buckling B^2 was found from equation (249) in the form

$$B^2 = N_u\sigma_a(\frac{\sigma_f}{\sigma_a}\nu - 1)/D$$

where N_u was the number of atoms of uranium per unit volume of oxide, D the diffusion coefficient and σ_a, σ_f the absorption and fission cross-sections of uranium for fast neutrons. Taking N_u appropriate to U_3O_8 of density 4.2 g/cm³ and

$$D = 3\tfrac{1}{3}\text{ cm}$$
$$\sigma_a = 0.1\ b \text{ (as quoted in Section 21.5)}$$
$$\frac{\sigma_f}{\sigma_a}\nu = 3$$

the value of B was found to be about $\frac{1}{40}$ cm^{-1}. The radius R was thus about 130 cm and the 'critical mass' of uranium some 40 *Te*.

The calculation was extended to find the reduction in critical mass when it was surrounded by an iron reflector 350 mm thick. The simple relation between B and R derived above was now replaced by the more complicated one obtained from solving equation (247) for the spherical mass of uranium and (68) for the reflector. The solutions were subject to the flux vanishing at the extrapolated outer boundary of the reflector and to continuity of flux and net current density across the interface between reflector and uranium. With an absorption cross-section for iron of 0.05 b the critical mass of uranium was reduced to about 12 Te.

Using up-to-date nuclear data, it is clear that criticality could not be achieved with a homogeneous mass of undiluted oxide of natural uranium. In fact, the fast neutron data of Appendix 8 suggest that less than 20 per cent of the absorptions in natural uranium give rise to fission whereas it seems to have been assumed above that all of them would do so. In consequence, the bracketed term in the expression for B^2 above would be strongly negative, making criticality impossible. To achieve criticality even in an infinite mass of oxide would require the concentration of uranium-235 to be increased about tenfold. To achieve criticality in the bare spherical mass as calculated above would require a further increase in this concentration by a factor of about two. The calculation, however, despite the paucity of nuclear data at the time it was made, is of some historical interest in providing an early example of the determination of critical size for a 'bare' and a 'reflected' reactor.

23.2 Criticality on thermal neutrons

With only natural uranium available for fission, a chain reaction maintained by fast neutrons, as in today's 'fast' reactor, is impossible, as was pointed out in the previous section. From equation (249) and the thermal cross-sections listed in Appendix 8, however, it is clear that criticality can be achieved in pure natural uranium if the neutrons emitted on fission can be sufficiently thermalised without undue loss. In order to thermalise them, it is necessary to introduce a material which meets the twin requirements of a high resonance escape probability and relatively low absorption of neutrons once thermalised. In modern power reactor parlance, a material which meets these requirements is termed a 'moderator' and a reactor which incorporates it is referred to as a 'thermal' reactor. From equation (214) the first of the two requirements of a moderator will be met by a high value of the mean logarithmic energy loss per collision, ξ, a low ratio of absorption to scattering cross-section at energies above thermal and a large volume relative to that of the uranium or uranium-bearing material, now generally referred to as the 'fuel'. The second requirement demands a low ratio of absorption to scattering cross-section at thermal energies but, in contrast to the first, a small volume relative to that of the fuel. It follows that the volume of moderator relative to that of the fuel will have some optimum value. To meet the first requirement a moderator can be expected to have a low mass number, and an obvious candidate is hydrogen, as has been used in the form of paraffin to achieve a source of thermal neutrons in experimental work described earlier. Modern power reactors may employ hydrogen as a moderator in the form of ordinary, or 'light', water (H_2O). The only other serious contenders are heavy water (D_2O) or graphite.

That part of a nuclear reactor which forms the critical mass is referred to as the 'core'. In general it is built up of a number of contiguous 'cells', each of which contains the same proportions of fuel, moderator, coolant, void and structural material and which are arranged in a regular array, or on a 'lattice'. If the core materials are intimately mixed, or separated within cells having a lattice pitch much less than that of the neutron mean free path, giving a 'homogeneous' reactor, then criticality is still not attainable on natural uranium even though the moderator has its optimum volume. As early as 1939, however, Fermi and L. Szilard had suggested the use of lattice cells on a pitch much larger than the neutron mean free path and in which fuel and moderator were segregated so that the local neutron flux within a cell would be depressed in the fuel relative to that in the moderator. Although this depression would enhance absorption of thermal neutrons in the moderator at the expense of that in the fuel, this disadvantage would be more than offset by the similar effect on resonance absorption, that is, by an increase in resonance escape probability. A reactor employing a lattice of this kind is referred to as a 'heterogeneous' reactor and should achieve criticality even on natural uranium fuel. In the interests of heat production at high temperatures, however, modern thermal reactors are no longer restricted to natural uranium fuel but use uranium which is slightly enriched in uranium-235.

Throughout the years 1940 and 1941, Fermi, Szilard, H. L. Anderson and E. P. Wigner had improved the accuracy of measurements on uranium cross-sections, on neutron capture and slowing-down in carbon, on resonance absorption in uranium-238 and of calculations on lattices. These are carried out on a single lattice cell as part of an infinite array so that neutron diffusion between cells need not be allowed for. The results provide average, or 'smeared', nuclear properties over the cell which can then be regarded as those of a homogeneous core. Experimental confirmation of these calculations was now required and was obtained as described in the section which follows.

23.3 The exponential experiment

Although a formula such as (251) can furnish a useful guide to the way in which the buckling depends on the nuclear properties of the material of a system, it cannot be relied upon for a sufficiently accurate value of the buckling, particularly when the information is to be applied to a project such as a nuclear power reactor. In such an application further confirmation is required and this can be provided by what is known as an 'exponential experiment'. It is carried out on a block having the same materials disposed in the same way as in the core of the proposed reactor. Being smaller than the reactor core itself, the block will be subcritical but, as already described in Section 22.4, a steady neutron flux can be set up within it by introducing a neutron source at one end. In fact, the arrangement is similar to that described in Section 17.4 for the measurement of diffusion length and a similar theory applies. In this case, however, the term $-\phi/L^2$ in the diffusion equation must be replaced by $+B^2\phi$ so that (178) becomes

$$\gamma_{mn}^2 = (\pi/a)^2(m^2 + n^2) - B^2 \tag{262}$$

and (179) can then be retained unaltered. As before, except near the boundaries, the fundamental term in equation (179), for which $m=n=1$, is found to be by far the most significant. The variation of neutron flux in the z-direction is thus given by the previous expression

$$e^{-\gamma_{11}z} - e^{-2\gamma_{11}b}\, e^{\gamma_{11}z}$$

or

$$e^{-\gamma_{11}z}\,[1 - e^{-2\gamma_{11}(b-z)}]$$

Provided, therefore, that z is sufficiently removed from b, the term in the square brackets is close to unity and the neutron flux falls off in an approximately exponential manner. It is this feature of the experiment which gives it its name. Comparison of this exponential variation with the measured one then enables an appropriate value of γ_{11} to be determined so that the required value of the buckling is found from the relation (262), which in this case becomes

$$\gamma_{11}{}^2 = 2(\pi/a)^2 - B^2$$

The theory of the exponential experiment was developed by Fermi in 1941. By this time Fermi and his co-workers were making use of a 'reproduction factor', now generally termed the 'infinite multiplication factor' and denoted by the symbol k_∞. In terms of the nomenclature used so far this factor is given by the relation

$$k_\infty = \nu\Sigma_f/\Sigma_a \qquad (263)$$

which, from (249), may be written

$$k_\infty = 1 + L^2B^2 \qquad (264)$$

It is thus seen to be the number of neutrons generated per neutron absorbed, provided that the reaction occurs in an infinite medium so that leakage of neutrons is absent. For criticality to be possible in a finite medium it is necessary to achieve a value of k_∞ greater than unity. By the end of 1941 Fermi and his team had devised a geometrical arrangement of sufficiently pure graphite and natural uranium to be able to measure a value of k_∞ equal to 0.87. By May 1942 the impurities had been reduced sufficiently to permit a value of 0.98 to be achieved.

There now followed an intensive programme of some thirty exponential experiments under the direction of Fermi and S. Allison. In the final form of these experiments, the lattice structure was of square cross-section having a side of 2.52 m and was 2.14 m in height. It was made up of thirty layers, each 105 mm thick. Half of these were of solid graphite whilst placed between each of them was a layer of graphite having a number of cavities, each of which contained uranium. In each uranium-bearing layer the cavities were uniformly spaced at a distance of 210 mm which thus became the pitch of the resulting cubic lattice. The uranium was in the form either of cast metal cylinders, 18 g/cm^3 in density and 57 mm in diameter by 57 mm long, or of pressed uranium oxide pseudo-spheres, 6.1 g/cm^3 in density and made by chamfering the corners of cylinders 82 mm in diameter by 82 mm long so that they were roughly spherical. The whole assembly stood on a graphite base 406 mm in height, the top layer of which contained a horizontal source channel. Two radium-beryllium fast neutron sources were placed in the channel, each roughly half-way from the centre to the edge of the assembly, so that the influence of the higher harmonics in the neutron flux distribution was diminished. In order to eliminate errors due to slow neutrons being scattered back into the assembly from the surroundings, the top and sides of the assembly were covered with cadmium sheet. To determine the axial variation in neutron flux, indium foils were positioned along the axis of the assembly and the induced 54-minute activity measured on Geiger–Müller counters. The activity was due to both thermal and indium resonance neutrons and was corrected to its saturation value.

The results obtained for B^2 were 45 and 59 $\times$ 10^{-6} cm^{-2} for uranium oxide in two different grades of graphite and 102 $\times$ 10^{-6} cm^{-2} for uranium metal in the higher grade of graphite. With diffusion lengths respectively equal to 26.7, 25.8 and 26.5 cm, the corresponding values of k_∞ were given by the relation (264) as 1.032, 1.039 and 1.067.

23.4 The first nuclear reactor

To check the nuclear properties of small reactors an alternative to the exponential experiment is the 'critical assembly'. Typically, it requires the reactor core to be set up without fuel but with a neutron source at or near the centre. At various points within the assembly instruments are located for measuring the neutron flux. Fuel is then loaded into the core in stages. At any particular stage, the ratio of the current flux level to that without fuel determines the 'neutron multiplication'. This is the ratio of the neutron flux actually achieved to that which would have been achieved on source neutrons alone, that is, in the absence of fission-generated neutrons. In an infinite medium and in the absence of neutrons generated by fission, the neutron flux due to a uniformly distributed neutron source of strength S will be

$$S/\Sigma_a$$

If neutrons generated by fission are now taken into account, the flux becomes, under subcritical conditions,

$$S/(\Sigma_a - \nu\Sigma_f)$$

The neutron multiplication is thus

$$\Sigma_a/(\Sigma_a - \nu\Sigma_f) = 1/(1 - k_\infty)$$

from the relation (263). In the case of a finite medium it will be necessary to allow for neutron leakage and it is possible to derive a more general result which does so.

The multiplication can be plotted against the mass of fuel loaded and its approach to infinity will then indicate the onset of criticality. More conveniently, the reciprocal of the multiplication may be plotted rather than the multiplication itself so that criticality is indicated as the curve obtained falls to zero. With careful siting of source and instrumentation the curve should take the form of a roughly straight line of negative slope so that the critical fuel loading can be deduced by extrapolation without the need to reach the critical state.

An alternative procedure is to start with an assembly, complete with fuel, which is well below the anticipated critical size and then to build it up in stages, taking neutron flux measurements at each stage, until criticality is approached. This is the procedure which was now adopted by Fermi and his associates following completion of the exponential experiments described in the previous section, not as an alternative to them, but as the final stage in constructing what was to be no less than the world's first nuclear reactor.

By this time Fermi and his team had moved to the University of Chicago and the site chosen for the reactor was a squash court under the west stand of the disused Stagg Athletic Field. The reactor itself was referred to at the time as a 'chain reacting pile' and was code-named CP (Chicago Physics) – 1. A group under W. H. Zinn took charge of the preparation of the materials and the construction, and a second group under V. C. Wilson looked after the controls and instrumentation. The base of the reactor was a grid of 150 × 100 mm timber, forming a square of 7.5 m side, and on this the graphite layers were stacked. The stack employed the same lattice geometry and the same three combinations of graphite and uranium as were used in the exponential experiments described in the previous section. It was planned that the central zone of the stack would contain the uranium metal cylinders and approximate in shape to a sphere some 3.3 m in diameter. This would be surrounded by a roughly spherical shell about 6.5 m in outside diameter formed of the higher quality graphite with uranium oxide pseudo-spheres. The outermost zone would then consist of uranium oxide pseudo-spheres in the lower quality graphite and approximate to a spherical shell some 7.8 m in outside diameter. The lower half of the stack was surrounded by vertical wooden sides erected on the four edges of the timber base and the space between the open box so formed and the graphite stack was filled with wood. In case it might prove necessary to evacuate the area, the reactor was constructed inside a tent of rubberised balloon fabric that could have been sealed. No neutron source was incorporated in the stack. Instead, reliance was placed on neutrons emitted as a result of the relatively low rate of spontaneous fission that occurs naturally in uranium. To monitor the neutron flux as the stack was built up, a boron trifluoride neutron counter was pushed along a horizontal slot at a height of about 1.1 m until it was just beyond the centre of the stack. Horizontal slots were also provided just below and above the centre of the stack for the insertion of control rods consisting of steel strip backed with cadmium sheet. A similar steel strip backed with boron steel sheet formed a safety rod which could be inserted in a further horizontal slot at the stack mid-height.

The count rate as measured had first to be corrected to eliminate that proportion of it due to increase in reactor size alone, the correction itself having to allow for departure of the graphite stack from a true spherical shape, particularly in the early stages of loading. The reciprocal of the count rate so corrected was then plotted against the height of the stack to arrive at the critical value.

The final layer of graphite was laid on 1 December 1942, to give an overall stack height of 6.2 m. This was well below the planned figure so that the stack eventually took the form of a sphere with the top sliced off. The final tests were made on the following day. Excitement mounted as the last control rod was gradually withdrawn until the point of predicted criticality had been reached. Then the neutron flux rose exponentially to record a power output of about 1 watt before the safety rods went in. The first nuclear reactor had successfully operated.

In the light of this achievement it may not be out of place to recall that, as far back as 1908, Soddy had held up in public a bottle containing 400 g of uranium oxide, proclaiming that therein 'lay asleep and waiting to be evolved the energy of about 900 tons of coal'. Much later, doubts set in so that, by 1933, a scientist of the eminence of Rutherford could assert that atomic energy could never be of any practical value. But by the afternoon of 2 December 1942, speculation was at an end and nuclear power was a reality.

Appendix 1

Some atomic constants and relationships

Faraday constant	(F)	96 487 coulombs/monovalent g atom
Charge-to-mass ratio of electron	(e/m)	1.7588×10^{8} coulombs/g
Electronic charge	(e)	1.6022×10^{-19} coulombs
Ratio of mass of hydrogen atom to mass of electron		1837
Avogadro number	(N_0)	6.0222×10^{23} amu/g
Mass of neutron		1.008665 amu
Energy–mass conversion factor		931.48 MeV/amu
Velocity of light	(c)	2.9979×10^{10} cm/sec
Energy of particle at fraction β of speed of light		$931.48\,[(1 - \beta^2)^{-\frac{1}{2}} - 1]$ Mev/amu
Boltzmann's constant		8.6171×10^{-5} eV/^{0}K
1 Curie		$= 3.7 \times 10^{10}$ disintegrations/sec
1 eV		$= 1.6022 \times 10^{-12}$ ergs

From *Tables of Physical and Chemical Constants* by G. W. C. Kaye and T. H. Laby, 14th edition, Longmans, Green & Co, London, New York and Toronto, 1973.

Appendix 2

Atomic properties of the naturally occurring stable elements

Element	Symbol	Atomic number Z	Atomic weight $\mathscr{A}$ (amu)	Mass number A	Fractional abundance (%)	Isotopic weight $\mathscr{M}$ (amu)
Hydrogen	H	1	1.0080	1 2	99.985 0.015	1.0078 2.0141
Helium	He	2	4.0026	3 4	0.00013 99.9999	3.0160 4.0026
Lithium	Li	3	6.941	6 7	7.42 92.58	6.0151 7.0160
Beryllium	Be	4	9.0122	9	100	9.0122
Boron	B	5	10.812	10 11	19.78 80.22	10.0129 11.0093
Carbon	C	6	12.0111	12 13	98.892 1.108	12.0000 13.0034
Nitrogen	N	7	14.0067	14 15	99.635 0.365	14.0031 15.0001
Oxygen	O	8	15.9994	16 17 18	99.759 0.037 0.204	15.9949 16.9991 17.9992
Fluorine	F	9	18.9984	19	100	18.9984
Neon	Ne	10	20.171	20 21 22	90.92 0.257 8.82	19.9924 20.9938 21.9914

Element	Symbol	Atomic number Z	Atomic weight $\mathscr{A}$ (amu)	Mass number A	Fractional abundance (%)	Isotopic weight $\mathscr{M}$ (amu)
Sodium	Na	11	22.9898	23	100	22.9898
Magnesium	Mg	12	24.312	24 25 26	78.60 10.11 11.29	23.9850 24.9858 25.9826
Aluminium	Al	13	26.9815	27	100	26.9815
Silicon	Si	14	28.089	28 29 30	92.18 4.71 3.12	27.9769 28.9765 29.9738
Phosphorus	P	15	30.9738	31	100	30.9738
Sulphur	S	16	32.06	32 33 34 36	95.0 0.760 4.22 0.014	31.9721 32.9715 33.9679 35.9671
Chlorine	Cl	17	35.449	35 37	75.79 24.20	34.9689 36.9659
Argon	A	18	39.9477	36 38 40	0.337 0.063 99.600	35.9675 37.9627 39.9624
Potassium	K	19	39.1422	39 41	93.22 6.77	38.9637 40.9618
Calcium	Ca	20	40.0783	40 42 43 44 46 48	96.97 0.64 0.145 2.06 0.0033 0.185	39.9626 41.9586 42.9588 43.9555 45.9537 47.9525
Scandium	Sc	21	44.9559	45	100	44.9559
Titanium	Ti	22	47.8798	46 47 48 49 50	7.99 7.32 73.99 5.46 5.25	45.9526 46.9518 47.9480 48.9479 49.9448

Element	Symbol	Atomic number Z	Atomic weight $\mathscr{A}$ (amu)	Mass number A	Fractional abundance (%)	Isotopic weight $\mathscr{M}$ (amu)
Vanadium	V	23	50.9415	51	99.75	50.9440
Chromium	Cr	24	51.9976	50 52 53 54	4.31 83.76 9.55 2.38	49.9461 51.9405 52.9407 53.9389
Manganese	Mn	25	54.9381	55	100	54.9381
Iron	Fe	26	55.8463	54 56 57 58	5.84 91.68 2.17 0.31	53.9396 55.9349 56.9354 57.9333
Cobalt	Co	27	58.9332	59	100	58.9332
Nickel	Ni	28	58.7047	58 60 61 62 64	67.76 26.16 1.25 3.66 1.16	57.9353 59.9308 60.9283 61.9297 63.9280
Copper	Cu	29	63.5470	63 65	69.1 30.9	62.9296 64.9278
Zinc	Zn	30	65.3801	64 66 67 68 70	48.89 27.81 4.11 18.56 0.62	63.9291 65.9261 66.9272 67.9249 69.9253
Gallium	Ga	31	69.7152	69 71	60.5 39.5	68.9256 70.9247
Germanium	Ge	32	72.6296	70 72 73 74 76	20.55 27.37 7.67 36.74 7.67	69.9243 71.9221 72.9235 73.9212 75.9214
Arsenic	As	33	74.9216	75	100	74.9216

Element	Symbol	Atomic number Z	Atomic weight $\mathscr{A}$ (amu)	Mass number A	Fractional abundance (%)	Isotopic weight $\mathscr{M}$ (amu)
Selenium	Se	34	78.9903	74	0.87	73.9225
				76	9.02	75.9192
				77	7.58	76.9199
				78	23.52	77.9173
				80	49.82	79.9165
				82	9.19	81.9167
Bromine	Br	35	79.9069	79	50.52	78.9183
				81	49.48	80.9163
Krypton	Kr	36	83.8040	78	0.354	77.9204
				80	2.27	79.9164
				82	11.56	81.9135
				83	11.55	82.9141
				84	56.90	83.9115
				86	17.37	85.9106
Rubidium	Rb	37	85.4681	85	72.15	84.9118
Strontium	Sr	38	87.6165	84	0.56	83.9134
				86	9.86	85.9093
				87	7.02	86.9093
				88	82.56	87.9056
Yttrium	Y	39	88.9059	89	100	88.9059
Zirconium	Zr	40	91.2237	90	51.46	89.9047
				91	11.23	90.9056
				92	17.11	91.9050
				94	17.40	93.9063
				96	2.80	95.9083
Niobium	Nb	41	92.9064	93	100	92.9064
Molybdenum	Mo	42	95.8863	92	15.86	91.9068
				94	9.12	93.9051
				95	15.70	94.9058
				96	16.50	95.9047
				97	9.45	96.9060
				98	23.75	97.9054
				100	9.62	99.9075

Element	Symbol	Atomic number Z	Atomic weight $\mathscr{A}$ (amu)	Mass number A	Fractional abundance (%)	Isotopic weight $\mathscr{M}$ (amu)
Ruthenium	Ru	44	101.0581	96	5.46	95.9076
				98	1.868	97.9053
				99	12.63	98.9059
				100	12.53	99.9042
				101	17.02	100.9056
				102	31.6	101.9043
				104	18.87	103.9054
Rhodium	Rh	45	102.9055	103	100	102.9055
Palladium	Pd	46	106.3563	102	0.96	101.9056
				104	10.97	103.9040
				105	22.2	104.9051
				106	27.3	105.9035
				108	26.7	107.9039
				110	11.8	109.9052
Silver	Ag	47	107.8779	107	51.35	106.9051
				109	48.65	108.9048
Cadmium	Cd	48	112.4341	106	1.22	105.9065
				108	0.88	107.9042
				110	12.39	109.9030
				111	12.75	110.9042
				112	24.07	111.9028
				114	28.86	113.9034
				116	7.58	115.9048
Indium	In	49	114.8193	113	4.23	112.9041
				115	95.77	114.9039
Tin	Sn	50	118.7411	112	0.95	111.9048
				114	0.65	113.9028
				115	0.34	114.9034
				116	14.24	115.9017
				117	7.57	116.9030
				118	24.01	117.9016
				119	8.58	118.9033
				120	32.97	119.9022
				122	4.71	121.9034
				124	5.98	123.9053

Element	Symbol	Atomic number Z	Atomic weight $\mathscr{A}$ (amu)	Mass number A	Fractional abundance (%)	Isotopic weight $\mathscr{M}$ (amu)
Antimony	Sb	51	121.7590	121	57.25	120.9038
				123	42.75	122.9042
Tellurium	Te	52	127.6395	120	0.089	119.9040
				122	2.46	121.9031
				124	4.61	123.9028
				125	6.99	124.9044
				126	18.71	125.9033
				128	31.79	127.9045
				130	34.49	129.9062
Iodine	I	53	126.9045	127	100	126.9045
Xenon	Xe	54	131.2451	124	0.096	123.9060
				126	0.090	125.9043
				128	1.919	127.9035
				129	26.44	128.9048
				130	4.08	129.9035
				131	21.18	130.9051
				132	26.89	131.9042
				134	10.4	133.9054
				136	8.87	135.9072
Caesium	Cs	55	132.9054	133	100	132.9054
Barium	Ba	56	137.3245	130	0.101	129.9063
				132	0.097	131.9051
				134	2.42	133.9046
				135	6.59	134.9055
				136	7.81	135.9043
				137	11.32	136.9055
				138	71.66	137.9050
Lanthanum	La	57	138.9053	139	99.911	138.9061
Cerium	Ce	58	140.1047	136	0.193	135.9070
				138	0.250	137.9058
				140	88.48	139.9054
				142	11.07	141.9091
Praseodymium	Pr	59	140.9076	141	100	140.9076

Element	Symbol	Atomic number Z	Atomic weight $\mathscr{A}$ (amu)	Mass number A	Fractional abundance (%)	Isotopic weight $\mathscr{M}$ (amu)
Neodymium	Nd	60	144.2234	142	27.13	141.9077
				143	12.20	142.9098
				145	8.29	144.9125
				146	17.18	145.9131
				148	5.72	147.9169
				150	5.60	149.9209
Samarium	Sm	62	150.3016	144	3.16	143.9120
				148	11.27	147.9148
				149	13.84	148.9172
				150	7.47	149.9172
				152	26.63	151.9198
				154	22.53	153.9223
Europium	Eu	63	151.9652	151	47.77	150.9198
				153	52.23	152.9212
Gadolinium	Gd	64	157.2662	154	2.15	153.9209
				155	14.7	154.9227
				156	20.47	155.9222
				157	15.68	156.9240
				158	24.9	157.9242
				160	21.9	159.9271
Terbium	Tb	65	158.9254	159	100	158.9254
Dysprosium	Dy	66	162.4943	156	0.0524	155.9239
				158	0.0902	157.9245
				160	2.294	159.9252
				161	18.88	160.9269
				162	25.53	161.9268
				163	24.97	162.9288
				164	28.18	163.9292
Holmium	Ho	67	164.9304	165	100	164.9304
Erbium	Er	68	167.2548	162	0.136	161.9287
				164	1.56	163.9293
				166	33.41	165.9303
				167	22.94	166.9321
				168	27.07	167.9324
				170	14.88	169.9356

Element	Symbol	Atomic number Z	Atomic weight $\mathscr{A}$ (amu)	Mass number A	Fractional abundance (%)	Isotopic weight $\mathscr{M}$ (amu)
Thulium	Tm	69	168.9342	169	100	168.9342
Ytterbium	Yb	70	173.0362	168 170 171 172 173 174 176	0.140 3.03 14.31 21.82 16.13 31.84 12.73	167.9342 169.9350 170.9364 171.9363 172.9381 173.9387 175.9427
Lutetium	Lu	71	174.9667	175	97.40	174.9406
Hafnium	Hf	72	178.4959	176 177 178 179 180	5.21 18.56 27.1 13.75 35.22	175.9416 176.9434 177.9439 178.9460 179.9469
Tantalum	Ta	73	180.9479	181	99.9877	180.9480
Tungsten	W	74	183.7222	180 182 183 184 186	0.135 26.4 14.4 30.6 28.4	179.9470 181.9483 182.9503 183.9510 185.9544
Rhenium	Re	75	186.2134	185	37.07	184.9531
Osmium	Os	76	190.3296	184 186 187 188 189 190 192	0.018 1.59 1.64 13.3 16.1 26.4 41.0	183.9528 185.9538 186.9558 187.9561 188.9583 189.9587 191.9615
Iridium	Ir	77	192.1921	191 193	38.5 61.5	190.9606 192.9630
Platinum	Pt	78	194.8503	192 194 195 196 198	0.78 32.9 33.8 25.2 7.19	191.9611 193.9627 194.9649 195.9650 197.9679

Element	Symbol	Atomic number Z	Atomic weight $\mathscr{A}$ (amu)	Mass number A	Fractional abundance (%)	Isotopic weight $\mathscr{M}$ (amu)
Gold	Au	79	196.9665	197	100	196.9665
Mercury	Hg	80	200.6090	196	0.146	195.9658
				198	10.02	197.9668
				199	16.84	198.9683
				200	23.13	199.9683
				201	13.22	200.9703
				202	29.80	201.9706
				204	6.85	203.9735
Thallium	Tl	81	204.3838	203	29.50	202.9724
				205	70.50	204.9744
Lead	Pb	82	208.2408	204	1.40	203.9730
				206	25.1	205.9745
				207	21.7	206.9759
				208	52.3	207.9767
Bismuth	Bi	83	208.9804	209	100	208.9804

All the figures in the above table have been taken from *Tables of Physical and Chemical Constants* by G. W. C. Kaye and T. H. Laby (14th edition, Longmans Green & Co, London, New York and Toronto, 1973). A few of the elements classed as stable do, in fact, have a radioactive isotope. In the above table, such radioactive isotopes have been excluded.

Appendix 3

The natural radioactive series

Element	Symbol	Atomic Number
Polonium	Po	84
Astatine	At	85
Radon	Rn	86
Francium	Fr	87
Radium	Ra	88
Actinium	Ac	89
Thorium	Th	90
Protactinium	Pa	91
Uranium	U	92

In the following tables, each isotope decays to the one listed immediately below it unless there is an arrow-headed line against it, in which case decay can only proceed as indicated by that line. When an isotope has two alternative modes of decay, the percentage of disintegrations that lead to the mode indicated by the right-hand line is given in the third column.

The uranium series

Radio-isotope		Radiation	Half-life
Uranium I	^{238}U	α	4.47×10^9 y
Uranium X_1	^{234}Th	β	24.1 d
Uranium X_2	^{234}Pa	β, γ (0.13% γ)	1.17 m
Uranium Z	^{234}Pa	β	6.70 h
Uranium II	^{234}U	α	2.454×10^5 y
Ionium	^{230}Th	α	7.54×10^4 y
Radium	^{226}Ra	α	1600 y
Radon	^{222}Rn	α	3.825 d
Radium A	^{218}Po	α (0.02% β)	3.05 m
(Astatine-218)	^{218}At	α (0.100% β)	1.600 s
Radium B	^{214}Pb	β	26.8 m
Radium C	^{214}Bi	β, γ (0.021% α)	19.9 m
(Radon-218)	^{218}Rn	α	5.40×10^{-4} s
Radium C′	^{214}Po	α	1.65×10^{-4} s
Radium C″	^{210}Tl	β	1.30 m
Radium D	^{210}Pb	β	22.20 y
Radium E	^{210}Bi	β (0.00013% α)	5.013 d
Radium F	^{210}Po	α	138.38 d
Radium E″	^{206}Tl	β	4.200 m
Lead	^{206}Pb	(stable)	–

The thorium series

Radio-isotope		Radiation	Half-life
Thorium	^{232}Th	α	1.45×10^{10} y
Mesothorium I	^{228}Ra	β	5.75 y
Mesothorium II	^{228}Ac	β, γ	6.13 h
Radiothorium	^{228}Th	α	1.9131 y
Thorium X	^{224}Ra	α	3.66 d
Thoron	^{220}Rn	α	55.6 s
Thorium A	^{216}Po	α	0.150 s
Thorium B	^{212}Pb	β	10.64 h
Thorium C	^{212}Bi	β (35.9% α)	60.6 m
Thorium C′	^{212}Po	α	3×10^{-7} s
Thorium C″	^{208}Tl	β, γ	3.053 m
Lead	^{208}Pb	(stable)	–

The actinium series

Radio-isotope		Radiation	Half-life
Actino-Uranium	^{235}U	α	7.038×10^8 y
Uranium Y	^{231}Th	β	1.0633 d
Protactinium	^{231}Pa	α	3.276×10^4 y
Actinium	^{227}Ac	β (1.38% α)	21.773 y
Radioactinium	^{227}Th	α	18.718 d
Actinium K	^{223}Fr	β (0.006% α)	21.80 m
(Astatine-219)	^{219}At	α (3% β)	54 s
(Bismuth-215)	^{215}Bi	β	7.4 m
Actinium X	^{223}Ra	α	11.43 d
Actinon	^{219}Rn	α	3.96 s
Actinium A	^{215}Po	α (0.0004% β)	0.00178 s
Actinium B	^{211}Pb	β	36.10 m
(Astatine-215)	^{215}At	α	1.0×10^{-4} s
Actinium C	^{211}Bi	α (0.273% β)	2.17 m
Actinium C′	^{211}Po	α	0.516 s
Actinium C″	^{207}Tl	β, γ	4.77 m
Lead	^{207}Pb	(stable)	–

The entries in these tables have been taken from 'The UK chemical nuclear data library: summary of the data available in the ENDF/B format' by B.S.J. Davies (CEGB Report RD/B/5095 N81, November 1981).

Appendix 4

Table of selected artificial radio-isotopes

Isotope	Radiation	Half-life	Isotope	Radiation	Half-life
^{3}H	β	12.3 y	^{51}Cr	γ	27.8 d
^{8}Li	β	0.84 s	^{56}Mn	β, γ	2.6 h
^{8}Be	α	2×10^{-16} s	^{59}Fe	β, γ	45 d
^{10}Be	β	2.5×10^{6} s	^{60}Co	β (I.T.)	10.5 m
^{12}Be	β, γ	0.011 s	^{60}Co	β, γ	5.3 y
^{14}C	β	5730 y	^{63}Ni	β	92 y
^{16}N	β, γ	7.2 s	^{65}Ni	β, γ	2.6 h
^{19}O	β, γ	29 s	^{64}Cu	β, β^{+}, γ	12.8 h
^{20}F	β, γ	11.4 s	^{66}Cu	β, γ	5.1 m
^{23}Ne	β, γ	37.6 s	^{65}Zn	β^{+}, γ	245 d
^{24}Na	γ	15 h	^{69}Zn	γ (I.T.)	13.8 h
^{27}Mg	β, γ	9.5 m	^{69}Zn	β	57 m
^{28}Al	β, γ	2.31 m	^{71}Zn	β, γ	2.4 m
^{31}Si	β	2.62 h	^{70}Ga	β	21.1 m
^{32}P	β	14.3 d	^{75}Ge	γ (I.T.)	48 s
^{35}S	β	88 d	^{80}Br	γ (I.T.)	4.38 h
^{37}S	β, γ	5.06 m	^{80}Br	β, β^{+}, γ	17.6 m
^{36}Cl	β	3.1×10^{5} y	^{79}Kr	γ (I.T.)	55 s
^{38}Cl	β, γ	37.3 m	^{79}Kr	β^{+}, γ	34.9 h
^{39}A	β	269 y	^{81}Kr	γ (I.T.)	13 s
^{41}A	β, γ	110 m	^{86}Rb	γ (I.T.)	1.04 m
^{42}K	β, γ	12.4 h	^{87}Sr	γ (I.T.)	2.83 h
^{45}Ca	β	165 d	^{94}Nb	γ (I.T.)	6.3 m
^{47}Ca	β, γ	4.53 d	^{94}Nb	β, γ	2×10^{4} y
^{49}Ca	β, γ	8.8 m	^{93}Mo	γ (I.T.)	6.9 h
^{46}Sc	γ (I.T.)	20 s	^{109}Pd	γ (I.T.)	4.7 m
^{46}Sc	β, γ	83.9 d	^{111}Pd	β, γ (I.T.)	5.5 h
^{47}Sc	β, γ	3.43 d	^{108}Ag	β, γ, β^{+}	2.42 m
^{49}Sc	β	57.5 m	^{107}Cd	β^{+}, γ	6.5 h
^{51}Ti	β, γ	5.8 m	^{111}Cd	γ (I.T.)	48.6 m
^{52}V	β, γ	3.76 m	^{114}In	γ (I.T.)	50 d

Isotope	Radiation	Half-life	Isotope	Radiation	Half-life
^{114}In	β, γ, β^+	72 s	^{185}W	β	75 d
^{121}Sn	β, γ (I.T.)	76 y	^{187}W	β, γ	23.9 h
^{122}Sb	γ (I.T.)	4.2 m	^{186}Re	β, γ	90 h
^{124}Sb	γ (I.T.)	24 m	^{188}Re	β, γ	16.7 h
^{124}Sb	β, γ (I.T.)	1.55 m	^{191}Os	γ (I.T.)	13 h
^{124}Sb	β, γ	60 d	^{191}Os	β, γ	15 d
^{121}Te	γ (I.T.)	154 d	^{193}Os	β, γ	31 h
^{123}Te	γ (I.T.)	117 d	^{192}Ir	β, γ (I.T.)	1.4 m
^{127}Xe	γ (I.T.)	75 s	^{192}Ir	β, γ	74.2 d
^{129}Xe	γ (I.T.)	8 d	^{194}Ir	β, γ	17.4 h
^{133}Ba	γ (I.T.)	38.9 h	^{195}Pt	γ (I.T.)	4.1 d
^{135}Ba	γ (I.T.)	28.7 h	^{197}Pt	β, γ (I.T.)	80 m
^{165}Dy	β, γ (I.T.)	1.26 m	^{197}Pt	β, γ	18 h
^{165}Dy	β, γ	2.34 h	^{199}Pt	β, γ	30 m
^{166}Ho	β, γ	26.9 h	^{198}Au	β, γ	2.7 d
^{169}Er	β	9.4 d	^{199}Au	β, γ	3.15 d
^{171}Er	β, γ	7.52 h	^{197}Hg	γ (I.T.)	24 h
^{170}Tm	β, γ	130 d	^{199}Hg	γ (I.T.)	43 m
^{171}Tm	β	1.92 y	^{203}Hg	β, γ	46.9 d
^{175}Yb	β, γ	4.2 d	^{205}Hg	β, γ	5.5 m
^{177}Yb	β, γ	1.9 h	^{204}Tl	β	3.8 y
^{176}Lu	β, γ (I.T.)	3.7 h	^{206}Tl	β	4.19 m
^{177}Lu	β, γ	6.7 d	^{207}Pb	γ (I.T.)	0.80 s
^{179}Hf	γ (I.T.)	18.6 s	^{209}Pb	β	3.3 h
^{180}Hf	γ (I.T.)	5.5 h	^{210}Bi	α, β (I.T.)	3×10^6 y
^{181}Hf	β, γ	42.5 d	^{210}Po	α	138.4 d
^{182}Ta	γ (I.T.)	16.5 m	^{233}Th	β, γ	22.2 m
^{182}Ta	β, γ	115 d	^{239}U	β, γ	23.5 m
^{183}W	γ (I.T.)	5.3 s	^{239}Np	β, γ	2.35 d
^{185}W	γ (I.T.)	96 s	^{239}Pu	α, γ	2.44×10^4 y

The abbreviation (I.T.) denotes an isomeric transition and the symbol β^+ a positron emission. The entries in this table have been taken from *Tables of Physical and Chemical Constants* by G. W. C. Kaye and T. H. Laby (14th edition, Longmans Green & Co., London, New York and Toronto, 1973).

Appendix 5

Thermal neutron cross-sections of some naturally occurring elements

Element	Atomic weight $\mathscr{A}$	Density ϱ (g/cm^3)	Microscopic cross-section (cm^2 × 10^{-24} or 'barns')		Macroscopic cross-section cm^{-1}	
			Scattering σ_s (average)	Absorption σ_a (at 0.025 eV)	Scattering Σ_s (average)	Absorption Σ_a (at 0.025 eV)
H	1.008	89.9 × 10^{-6} (at N.T.P.)	45.3	0.332	0.00243	0.0000178
He	4.003	178 × 10^{-6} (at N.T.P.)	0.72	0.0069	0.0000194	1.85 × 10^{-7}
Li	6.941	0.53	1.03	70	0.0474	3.23
Be	9.012	1.85	6.0	0.0095	0.74	0.00117
B	10.812	2.47	4.46	759	0.61	104
C	12.011	1.75 (reactor graphite)	4.69	0.00340	0.411	0.000298
N	14.007	1251 × 10^{-6} (at N.T.P.)	11.0	1.90	0.00059	0.000102
O	15.999	1429 × 10^{-6} (at N.T.P.)	3.85	0.000322	0.000207	1.73 × 10^{-8}
F	18.998	1695 × 10^{-6} (at N.T.P.)	3.82	0.0100	0.000205	5.4 × 10^{-7}
Na	22.990	0.97	3.31	0.53	0.084	0.0135
Mg	24.312	1.74	3.47	0.064	0.149	0.00277
Al	26.982	2.70	1.41	0.229	0.085	0.0138
Si	28.089	2.33	2.10	0.160	0.105	0.0080
Cl	35.449	3214 × 10^{-6} (at N.T.P.)	17.1	32.0	0.00094	0.00175
K	39.142	0.86	2.18	2.07	0.0289	0.0275
Ca	40.078	1.53	3.00	0.437	0.069	0.0101
Ti	47.880	4.51	4.52	5.7	0.256	0.325
V	50.94	6.1	5.0	5.0	0.360	0.364
Cr	52.00	7.2	4.10	3.10	0.342	0.258
Mn	54.94	7.5	1.72	13.4	0.141	1.10
Fe	55.85	7.9	11.3	2.55	0.96	0.217
Co	58.93	8.8	6.7	37.2	0.60	3.35
Ni	58.70	8.9	17.4	4.42	1.59	0.404
Cu	63.55	8.9	6.7	3.82	0.57	0.323

Element	Atomic weight $\mathscr{A}$	Density ϱ (g/cm^3)	Microscopic cross-section (cm^2 × 10^{-24} or 'barns') Scattering σ_s (average)	Microscopic cross-section Absorption σ_a (at 0.025 eV)	Macroscopic cross-section cm^{-1} Scattering Σ_s (average)	Macroscopic cross-section cm^{-1} Absorption Σ_a (at 0.025 eV)
Ga	69.72	5.9	6.9	2.98	0.353	0.152
Zr	91.22	6.5	6.2	0.183	0.268	0.0079
Nb	92.91	8.6	5.7	1.15	0.315	0.064
Mo	95.89	10.2	4.47	2.70	0.287	0.173
Ag	107.88	10.5	3.69	64	0.217	3.77
Cd	112.43	8.6	10.1	2450 (not 1/v)	0.469	113 (not 1/v)
Sn	118.74	7.3	4.79	0.63	0.177	0.0233
Eu	151.97	5.2	3.82	4700 (not 1/v)	0.080	98 (not 1/v)
Gd	157.27	7.9	142	48900 (not 1/v)	4.29	1480 (not 1/v)
Hf	178.50	13.3	8.6	103	0.385	4.60
Ta	180.95	16.7	5.1	21.0	0.283	1.17
W	183.72	19.3	4.99	18.4	0.315	1.16
Pt	194.85	21.5	12.0	8.0	0.80	0.53
Au	196.97	19.3	7.7	99	0.455	5.8
Pb	208.24	11.3	8.6	0.170	0.281	0.0056

The densities of gases have been taken from 'Kempe's Engineers Year-Book 1983', Ed. J. P. Quayle, (Morgan-Grampian Book Publishing Co. Ltd., London). The density of carbon as graphite has been obtained, by kind permission of the U.K.A.E.A., from their document 'Properties of Substances (in SI units)' (Atomic Energy Technical Data Sheets, UDC53 part 3, sheet 2, October 1980). All other densities have been taken from 'Tables of Physical and Chemical Constants' by G. W. C. Kaye and T. H. Laby (14th. edition, Longmans Green & Co., London, New York and Toronto, 1973) and are quoted at 20°C. The cross-sections have been taken from the U.K. Nuclear Data Library (U.K.A.E.A. Winfrith, 1981 edition). In the case of reactor graphite, it may be necessary to allow for the presence of elements other than carbon as impurities and the effect of irradiation on them.

Appendix 6

Neutron diffusion and slowing-down data for selected materials

Material	Density (g/cm^3)	L (cm)	Σ_a (cm^{-1})	D (cm)	L_s^* (cm)	β (infinite slab)
Water	1.00	2.7	0.019	0.14	5.17	0.81
Paraffin wax	0.89	2.1	0.025	0.11	4.2	0.81
Heavy water (pure)	1.10	94	0.000095	0.84	10.8	0.97
Beryllium	1.85	20.8	0.00116	0.50	8.56	0.91
Graphite	1.67	52	0.00031	0.85	17.3	0.94
Iron	7.86	1.27	0.21	0.34	18.2	0.30
Lead	11.35	12.18	0.0062	0.92	77.5	0.74
Concrete (1.0 wt.% hydrogen)	2.3	8.0	0.0095	0.60	10.0	0.74

* from fission source energy to 1.46 eV.

All results are quoted for a temperature of 20°C. Results for Σ_a and D are average values for thermal neutrons. In the case of concrete, Σ_a has been obtained by applying cross-sections from the UK Nuclear Data Library (U.K.A.E.A. Winfrith, 1981 edition) to a concrete of the given density and hydrogen content as specified in 'Radiation Shielding' by B. T. Price, C. C. Horton and K. T. Spinney (Pergamon Press, London, New York and Paris, 1957), and the result used to derive L. Otherwise, values of Σ_a and β are derived from the remaining data, which have been taken from 'Tables of Physical and Chemical Constants' by G. W. C. Kaye and T. H. Laby (14th. edition, Longmans Green & Co., London, New York and Toronto, 1973).

Appendix 7

Selected ^{235}U fission product decay chains

In the following tables, each isotope proceeds by β-decay to the one listed immediately below it unless there is an arrow-headed line against it, in which case decay can only proceed as indicated by that line. When an isotope has two alternative modes of decay, the percentage of disintegrations that lead to the mode indicated by the right-hand line is given in the 'Half-life' column. For any particular mass number, the yield quoted is a 'chain yield', i.e. the percentage of fissions that yield the stable isotope(s) of that mass number, whether the parent fission fragment is the first member of the chain or not.

Isotope	Half-life
Mass no 83 Yield 0.53%	
Zn	0.08 s
Ga	0.31 s
Ge	1.9 s
As	13.3 s (0.0062%)
Se	1.17 m
Se	22.5 m
Br	2.39 h (99.975%)
Kr	1.83 h
Kr	(stable)

Isotope	Half-life
Mass no 88 Yield 3.60%	
Ge	0.14 s
As	0.13 s
Se	1.5 s
Br	16.3 s
Kr	2.84 h
Rb	17.8 m
Sr	(stable)

Isotope	Half-life
Mass no 129 Yield 0.612%	
Cd	0.34 s
In	2.5 s
In	0.99 s (50%)
Sn	2.50 m
Sn	7.50 m
Sb	4.31 h (16.6%)
Te	33.6 d (35%)
Te	1.16 h
I	1.6 × 10^7 y
Xe	(stable)

Isotope	Half-life
Mass no 130 Yield 1.57%	
I	9.0 m (83%)
I	12.36 h
Xe	(stable)
Cd	0.52 s
In	0.50 s
Sn	3.8 m
Sb	6.30 m
Sn	1.7 m
Sb	40.0 m
Te	(stable)

Isotope	Half-life
Mass no 139 Yield 6.65%	
Sb	0.17 s
Te	0.42 s
I	2.4 s
Xe	39.5 s
Cs	9.40 m
Ba	1.38 h
La	(stable)

Isotope	Half-life
Mass no 140 Yield 6.31%	
Te	0.075 s
I	0.86 s
Xe	13.6 s
Cs	1.06 m
Ba	12.74 d
La	1.68 h
Ce	(stable)

Isotope	Half-life	Isotope	Half-life	Isotope	Half-life
Mass no 127 yield 0.115%		Mass no 131 Yield 2.78%		Mass no 141 Yield 5.81%	
Ag	0.21 s	Cd	0.12 s	Te	0.24 s
Cd	1.06 s (50%)	In	0.30 s	I	0.40 s
In	127 m	Sn	50.0 s	Xe	1.72 s
In	2.0 s	Sn	39.0 s	Cs	24.94 s
Sn	2.10 h	Sb	23.0 m (6.8%)	Ba	18.27 m
Sn	41.3 m	Te	1.25 d (77.8%)	La	3.93 h
Sb	3.89 d (17.4%)	Te	25.0 m	Ce	32.5 d
Te	109.0 d (2.4%)	I	8.04 d (1.1%)	Pr	(stable)
Te	9.35 h	Xe	11.9 d		
I	(stable)	Xe	(stable)		

Decay chains and half-lives taken from 'A revised fission product decay data file in ENDF/B-IV format' by A. Tobias and B. S. J. Davies (CEGB Report RD/B/N4942, November 1980). Yields taken from 'Fission-product yields from neutron-induced fission', by E. A. C. Crouch (Atomic Data and Nuclear Data Tables, Vol. 19, No. 5, Academic Press, New York & London, May 1977).

Appendix 8

Nuclear properties of some fissile isotopes

Isotope	Thermal Neutrons			Fast Neutrons		
	ν	σ_f (b)	σ_c (b)	ν	σ_f (b)	σ_c (b)
^{235}U	2.42	579	100	2.45	1.98	0.533
^{238}U	–	–	2.73	2.76	0.043	0.295
^{239}Pu	2.88	742	266	2.94	1.795	0.533
^{240}Pu	2.95	0.0077	281	3.145	0.350	0.607
^{241}Pu	2.96	1031	356	2.945	2.63	0.599

Data for thermal neutrons are quoted at 2200 m/s and are taken from the U.K. Nuclear Data Library (U.K.A.E.A. Winfrith, 1981 edition). Data for fast neutrons are spectrally averaged values for a sodium-cooled fast reactor with mixed oxide fuel and are taken from 'Influence of fast reactor design on system cost and uranium requirements', by C. E. Iliffe, A. Green, R. Herrick and Mrs J. M. Ball, Paper No 50, British Nuclear Energy Society Conference on Optimisation of Sodium-Cooled Fast Reactors, London, 28 November–1 December 1977.

References

1 General

S. Glasstone, *Sourcebook on Atomic Energy*, Macmillan, London, 1950.

S. Glasstone and M. C. Edlund *The Elements of Nuclear Reactor Theory*, Macmillan, London, 1952.

J. B. Hoag, *Electron and Nuclear Physics*, 3rd edn, d Van Nostrand, New York, 1952.

I. Kaplan, *Nuclear Physics*, Addison-Wesley, Reading, Mass, USA, 1956.

B. T. Price, C. C. Horton and K. T. Spinney, *Radiation Shielding*, Pergamon Press, London, New York & Paris, 1957.

Sir E. Rutherford, J. Chadwick and C. D. Ellis, *Radiations from Radioactive Substances*, Cambridge University Press, 1930.

F. Soddy, *The Story of Atomic Energy*, Nova Atlantis Publishing, London, 1949.

S. Tolansky, *Introduction to Atomic Physics*, Longmans, Green, London, New York, and Toronto, 1949.

M. R. Wehr and J. A. Richards Jr, *Physics of the Atom*, Addison-Wesley, Reading, Mass, USA, 1960.

A. M. Weinberg and E. P. Wigner, *The Physical Theory of Neutron Chain Reactors*, University of Chicago Press, 1958.

2 Particular

Chapter 6

E. Rutherford, 'Retardation of the α-particle from radium in passing through matter', *The London, Edinburgh and Dublin Philosophical Magazine & Journal of Science*, 6th Series, **12**, August 1906, pp. 134–46.

E. Rutherford, 'The mass & velocity of the α-particles expelled from radium and actinium', *The London, Edinburgh and Dublin Philosophical Magazine & Journal of Science*, 6th Series, **12**, Sept. 1906, pp. 348–71.

Chapter 8

P. M. S. Blackett, 'On the analysis of α-ray photographs', *Proceedings of the Royal Society*, Series A, **102**, 1922, p. 294.

J. D. Cockroft and E. T. S. Walton, 'Experiments with high velocity positive ions. – (I) Further developments in the method of obtaining high velocity positive ions', *Proceedings of the Royal Society of London*, Series A, **136**, June 1932, pp. 619–30.

J. D. Cockroft and E. T. S. Walton, 'Experiments with high velocity positive ions. II – The disintegration of elements by high velocity protons', *Proceedings of the Royal Society of London*, Series A, **137**, July 1932, pp. 229–42.

P. I. Dee and E. T. S. Walton, 'A photographic investigation of the transmutation of lithium and boron by protons and of lithium by ions of the heavy isotope of hydrogen', *Proceedings of the Royal Society of London*, Series A, **141**, Sept. 1933, pp. 733–40.

Chapter 9

N. Feather, 'The collisions of neutrons with nitrogen nuclei', *Proceedings of the Royal Society of London*, Series A, **136**, June 1932, pp. 709–27.

J. Chadwick, 'The existence of a neutron', *Proceedings of the Royal Society of London*, Series A, **136**, June 1932, pp. 692–708.

J. Chadwick, 'The neutron', *Proceedings of the Royal Society of London*, Series A, **142**, Oct. 1933, pp. 1–25.

Chapter 10

E. Fermi, E. Amaldi, O. D'Agostino, F. Rasetti and E. Segré, 'Artificial radioactivity produced by neutron bombardment', *Proceedings of the Royal Society of London*, Series A, **146**, 1934, pp. 483–500.

Chapter 11

N. Bohr, 'Neutron capture and nuclear constitution', *Nature*, **137**, 29th Feb. 1936, pp. 344–8.

Chapter 12

J. R. Dunning, 'The emission and scattering of neutrons', *The Physical Review*, **45**, 1st May 1934, pp. 586–600.

J. R. Dunning, G. B. Pegram, G. A. Fink and D. P. Mitchell, 'Interaction of neutrons with matter', *The Physical Review*, **48**, 1st Aug. 1935, pp. 265–80.

L. Cranberg, R. B. Day, L. Rosen, R. F. Taschek and M. Walt, 'Techniques for measuring elastic and non-elastic neutron cross-sections', *Progress in Nuclear Energy, Series I, Physics & Mathematics, Vol 1*, pp. 107–60, Pergamon Press, London, 1956.

D. J. Hughes and R. S. Carter, 'Neutron cross-section – Angular distributions', US Report BNL-400, June 1956.

Chapter 13

E. Amaldi, O. D'Agostino, E. Fermi, B. Pontecorvo, F. Rasetti and E. Segré, 'Artificial radioactivity produced by neutron bombardment – II', *Proceedings of the Royal Society of London*, Series A, **149**, 1935, pp. 522–58.

J. Chadwick and M. Goldhaber, 'Disintegration by slow neutrons', *Nature*, **135**, 12th Jan. 1935, p. 65.

H. J. Taylor and M. Goldhaber, 'Detection of nuclear disintegration in a photographic emulsion', *Nature*, **135**, 2nd March 1935, p. 341.

J. R. Dunning, G. B. Pegram, G. A. Fink and D. P. Mitchell, (as in Chapter 12)

J. R. Dunning, G. B. Pegram, G. A. Fink, D. P. Mitchell and E. Segré, 'Velocity of slow neutrons by mechanical velocity selector', *The Physical Review*, **48**, 15 Oct. 1935, p. 704.

G. A. Fink, 'The production and absorption of thermal energy neutrons', *The Physical Review*, **50**, 8, 15 Oct. 1936, pp. 738–47.

A. C. G. Mitchell, E. J. Murphy and L. M. Langer, 'Continuation of work on scattering of slow neutrons', *The Physical Review*, **49**, 1st March 1936, p. 400.

A. C. G. Mitchell, E. J. Murphy and M. D. Whitaker, 'Selective scattering of slow neutrons', *The Physical Review*, **49**, 1st March 1936, p. 401.

A. C. G. Mitchell, E. J. Murphy and M. D. Whitaker, 'Scattering of slow neutrons II', *The Physical Review*, **50**, 15 July 1936, pp. 133–7.

M. Goldhaber and G. H. Briggs, 'Scattering of slow neutrons', *Proceedings of the Royal Society of London*, Series A, **162**, A908, 1st Sept. 1937, pp. 127–43.

H. A. Bethe, 'Theory of disintegration of nuclei by neutrons', *The Physical Review*, **47**, 15th May 1935, pp. 747–59.

F. Rasetti, E. Segré, G. Fink, J. R. Dunning and G. B. Pegram, 'On the absorption law for slow neutrons', *The Physical Review*, **49**, 1st Jan. 1936, p. 104.

F. Rasetti, D. P. Mitchell, G. A. Fink and G. B. Pegram, 'On the absorption of slow neutrons in boron', *The Physical Review*, **49**, 15th May 1936, p. 777.

H. H. Goldsmith and F. Rasetti, 'On the resonance capture of slow neutrons', *The Physical Review*, **50**, 15 Aug. 1936, pp. 328–31.

Chapter 14

R. Peierls, 'Critical conditions in neutron multiplication', *Proceedings of the Cambridge Philosophical Society*, **35**, 1939, pp. 610–15.

A. H. Wilson, 'A comparison of the approximate methods of calculating the critical size of a sphere', Birmingham Report MS 89, 1943.

E. Fermi, 'On the motion of neutrons in hydrogenous substances', *Ricerca Scientifica*, VII – 2, 13 (1936).

Chapter 15

A. H. Wilson, 'The effect of anisotropic scattering on the multiplication in a sphere', Birmingham Report MS – 113, 21st July 1944.

E. Fermi, (as in Chapter 14)

Chapter 16

E. Fermi, (as in Chapter 14)

A. H. Wilson, 'The general properties of the transport equation and its use in determining the multiplication in bodies having spherical symmetry', Birmingham Report MS 105, 13th Jan. 1944.

R. Peierls, (as in Chapter 14)

K. M. Case, F. de Hoffmann and G. Placzek, 'Introduction to the theory of neutron diffusion', (Vol 1) AEC Report, Los Alamos Scientific Laboratory, June 1953.

E. Amaldi and E. Fermi, 'On the absorption and the diffusion of slow neutrons', *The Physical Review*, **50**, 15th Nov. 1936, pp. 899–928.

Chapter 17

E. Fermi, (as in Chapter 14)

E. Amaldi and E. Fermi, (as in Chapter 16)

H. A. Bethe, 'Nuclear physics B. Nuclear dynamics, theoretical', *Reviews of Modern Physics*, **9**, April 1937, 2, pp. 71–244.

H. L. Anderson and E. Fermi, 'Production and absorption of slow neutrons by carbon', Report A-21, Columbia University, New York, 25th Sept. 1940.

Chapter 18

E. Amaldi and E. Fermi, (as in Chapter 16)

G. Placzek, 'On the theory of the slowing down of neutrons in heavy substances', *The Physical Review*, **69**, 1st May 1946, pp. 423–38.

G. Placzek, 'Slowing down of neutrons by heavy nuclei', *The Physical Review*, **55**, 1st June 1939, p. 1130.

Chapter 19

P. B. Moon and J. R. Tillman, 'Evidence on the velocities of slow neutrons', *Nature*, **135**, 1st June 1935, p. 904.

J. R. Tillman and P. B. Moon, 'Selective absorption of slow neutrons', *Nature*, **136**, 13th July 1935, pp. 66–7.

E. Amaldi and E. Fermi, (as in Chapter 16)

H. H. Goldsmith and F. Rasetti, (as in Chapter 13)

G. Placzek, (as 1946 in Chapter 18)

H. A. Bethe and G. Placzek, 'Resonance effects in nuclear processes', *The Physical Review*, **51**, 15 March 1937, pp. 450–84.

Chapter 20

H. L. Anderson and E. Fermi, (as in Chapter 17)

E. Fermi (as in Chapter 14)

E. Amaldi and E. Fermi, (as in Chapter 16)

R. Ehrlich and H. Hurwitz Jun., 'Multigroup methods for neutron diffusion problems', *Nucleonics*, **12**, 2, Feb. 1954, pp. 23–30.

C. H. Westcott, 'Some calculations on the diffusion of slow neutrons in hydrogenous media', *Proceedings of the Cambridge Philosophical Society*, **33**, Part I, Jan. 1937, pp. 122–36.

Chapter 21

L. Meitner, O. Hahn, and F. Strassmann, 'Über die Umwandlungsreihen des Urans, die durch Neutronenbestrahlung erzeugt werden', *Zeitschrift für Physik*, **106**, 1937, pp. 249–70.

L. A. Turner, 'Nuclear fission', *Reviews of Modern Physics*, **12**, Jan. 1940, 1, pp. 1–29.

O. Hahn and F. Strassmann, 'Über die Entstehung von Radiumisotopen aus Uran durch Bestrahlen mit schnellen und verlangsamten Neutronen', *Die Naturwissenschaften*, **26**, 18th Nov. 1938, pp. 755–6.

O. Hahn and F. Strassmann, 'Über den Nachweis und das Verhalten der bei der Bestrahlung des Urans mittels Neutronen entstehenden Erdalkalimetalle', *Die Naturwissenschaften*, **27**, 6th Jan. 1939, pp. 11–15.

O. Hahn and F. Strassmann, 'Nachweis der Entstehung aktiver Bariumisotope aus Uran und Thorium durch Neutronenbestrahlung; Nachweis weiterer aktiver Bruchstücke bei der Uranspaltung', *Die Naturwissenschaften*, **27**, 10 Feb. 1939, pp. 89–95.

L. Meitner and O. R. Frisch, 'Disintegration of uranium by neutrons; a new type of nuclear reaction', *Nature*, **143**, 11th Feb. 1939, pp. 239–40.

O. R. Frisch, 'Physical evidence for the division of heavy nuclei under neutron bombardment', *Nature*, **143**, 18th Feb. 1939, p. 276.

L. Meitner and O. R. Frisch, 'Products of the fission of the uranium nucleus', *Nature*, **143**, 18th March 1939, pp. 471–2.

F. A. Heyn, A. H. W. Aten Jun., and C. J. Bakker, 'Transmutation of uranium and thorium by neutrons', *Nature*, **143**, 25th March 1939, pp. 516–17.

E. Bretscher and L. G. Cook, 'Transmutations of uranium and thorium nuclei by neutrons', *Nature*, **143**, 1st April 1939, pp. 559–60.

R. Abelson, 'The identification of some of the products of uranium cleavage', *The Physical Review*, **55**, 1st May 1939, pp. 876–7.

R. W. Dodson and R. D. Fowler, 'Radioactive halogens produced by the neutron bombardment of uranium and thorium', *The Physical Review*, **55**, 1st May 1939, p. 880.

H. L. Anderson, E. Fermi and A. V. Grosse, 'Branching ratios in the fission of uranium (235)', *The Physical Review*, **59**, 1st Jan. 1941, pp. 52–6.

H. von Halban Jun., L. Kowarski and P. Savitch, 'Sur la capture simple des neutrons thermiques et des neutrons de resonance par l'uranium', *Comptes Rendus des Séances de l'Académie des Sciences*, **208**, 1939, pp. 1396–8.

H. L. Anderson, E. T. Booth, J. R. Dunning, E. Fermi, G. N. Glasoe and F. G. Slack, 'The fission of uranium', *The Physical Review*, **55**, 1st March 1939, pp. 511–12.

Chapter 22

H. von Halban Jun., F. Joliot and L. Kowarski, 'Liberation of neutrons in the nuclear explosion of uranium', *Nature*, **143**, 18th March 1939, pp. 470–1.

H. von Halban Jun., F. Joliot and L. Kowarski, 'Number of neutrons liberated in the nuclear fission of uranium', *Nature*, **143**, 22nd April 1939, p. 680.

H. von Halban J., F. Joliot and L. Kowarski, 'Energy of neutrons liberated in the nuclear fission of uranium induced by thermal neutrons', *Nature*, **143**, 3rd June 1939, p. 939.
R. Peierls, (as in Chapter 14)
A. H. Wilson, (as in Chapter 14)
A. H. Wilson, (as in Chapter 15)

Chapter 23

F. Perrin, 'Calcul relatif aux conditions éventuelles de transmutation en chaîne de l'uranium', *Comptes Rendus des Séances de l'Académie des Sciences*, **208**, May 1939, pp. 1394–6.
H. D. Smyth, 'Atomic energy for military purposes', US Government Printing Office, 1945.
Newsletter '40 years on', *Nuclear Energy*, 21, 6, December 1982, p. 346.
E. Fermi, 'Experimental production of a divergent chain reaction', USAEC Report AECD-3269, 4th January 1952.
M. Gowing, 'Principalities and nuclear power: the origins of reactor systems', *The Nuclear Engineer*, 23, 3, May/June 1982, pp. 70–6.

Index

Name index

Subject index